THE INFINITE HORIZON

THE INFINITE HORIZON

RESOLVING MANKIND'S COSMIC DEBATE

JOHN DAYTON

XULON PRESS

Xulon Press
2301 Lucien Way #415
Maitland, FL 32751
407.339.4217
www.xulonpress.com

© 2021 by John Dayton

All rights reserved solely by the author. The author guarantees all contents are original and do not infringe upon the legal rights of any other person or work. No part of this book may be reproduced in any form without the permission of the author.

Due to the changing nature of the Internet, if there are any web addresses, links, or URLs included in this manuscript, these may have been altered and may no longer be accessible. The views and opinions shared in this book belong solely to the author and do not necessarily reflect those of the publisher. The publisher therefore disclaims responsibility for the views or opinions expressed within the work.

Unless otherwise indicated, Scripture quotations taken from the New American Standard Bible (NASB). Copyright © 1960, 1962, 1963, 1968, 1971, 1972, 1973, 1975, 1977, 1995 by The Lockman Foundation. Used by permission. All rights reserved.

Paperback ISBN-13: 978-1-6628-2164-6
Ebook ISBN-13: 978-1-6628-2165-3

Dedication

My sister and I grew up in a loving home of modest means and cherished our parents' love and encouragement. Their endless sacrifice to nurture our young lives and teach us moral precepts helped shape us into successful professionals who desired to be like them. My father perished in an industrial accident when he was fifty-six, and my sister Judy lost her battle with cancer at fifty-nine. My mother Emma went on to live a full life, apart from her loving husband John, and passed into glory in 2014. I miss them and hold on to our family memories.

Despite my feelings of gratitude toward my parents, I have an overwhelming desire to dedicate this book to my loving wife Christine, whom I cherish and dearly love. You are the spark in my life, the spring in my step, and you have inspired and supported me to dream big and reach for the heavens. You have given us two lovely daughters, Sarah and Rebecca, who have been our central focus for many years. They have brought us great joy in life. Your hard work and dedication to our family and your profession as a teacher have been greatly appreciated. You have helped mentor our girls into successful relationships and careers. I've enjoyed our new role and time spent together mentoring our grandchildren and watching our family grow. By God's grace, I pray we have many more wonderful years together.

"In the beginning God created the heavens and the earth. And the earth was formless and void, and darkness was over the surface of the deep: and the Spirit of God was moving over the surface of the waters" (Gen. 1:1-2).

Author's Note

THIS BOOK IS as much about the evolution of mankind's ability to separate facts from desired outcomes as it is about discerning the truth concerning the creation of the universe. Throughout recorded history, scientific observations have not always been accurately interpreted or reassessed following the discovery of new evidence. This leads us to wonder if our current thinking concerning the Big Bang creation theory is entirely accurate.

Some of you may find that sections of this book are a bit complex and technical. In doing my research on this subject, I realized we needed to get "into the weeds" and lay out enough details to assess this subject properly and ensure we have constructed a credible case based on factual evidence instead of postulating about a desired outcome.

I've tried to make this book all-encompassing about the origin of the universe. To properly analyze this subject, one must discuss the related topics to ensure a factual solution is found that holds up over a wide range of topics. In doing so, I've chosen to make some chapters very brief by only discussing sufficient information to explore a particular area about the main objective for this book, which is finding the truth about how the infinite horizon came into existence.

Author's Note

I've always been fascinated with cosmology and have independently studied this subject over the years. Some may opine that the lack of a formal education in this area should disqualify me from expounding about the creation of the universe. However, I believe my independent research about this subject and my background education in nuclear physics has given me sufficient knowledge about this subject matter to render factual and accurate conclusions. I also believe my engineering background and troubleshooting experience has better prepared me to analyze and decipher the truth about our universe.

This book attempts to portray a factually based, analytically derived postulation about the universe. If any of the facts contained within this publication are found to be inaccurate or incomplete, then different conclusions could be derived. As outlined in this book, there have been numerous changes about the perceived reality surrounding this subject. The conclusions in this book are based off of factual evidence and analysis. I've also attempted to remove all bios and speculation in this analysis. As additional information about our universe is gathered and properly analyzed, we may find affirmation for the conclusions in this book or reason to reach a different outcome. Time and research will tell.

I want to thank those of you who have helped me to refine this document. Your assistance in aiding me to achieve my long-term goal of sharing my research about the cosmos and expressing it in book form has been invaluable!

Table of Contents

Chapter 1: The Mysteries of the Universe 1
Chapter 2: The Evolution of the Great Creation 7
Chapter 3: How Do We Know? 35
Chapter 4: The Elephant in the Room 39
Chapter 5: What's the Matter? 53
Chapter 6: The Gravity of the Situation 61
Chapter 7: How Really Big Was That Universe? 73
Chapter 8: Dark Matter Is No Matter at All 77
Chapter 9: Busting the Big Bang 83

Illustrations 107

Chapter 10: Just the Facts, Ma'am! 127
Chapter 11: What's Beyond Our View? 137
Chapter 12: Six Days or 6 Days? 141
Chapter 13: Waters of the Deep 155
Chapter 14: Is Creation Possible? 161
Chapter 15: How Far Back Does Creation Go? 165
Chapter 16: Creation vs. Evolution 171
Chapter 17: The Progressive Creation of Earth 177
Chapter 18: The Link Between Science and Religion 189
Chapter 19: Theories Derived 195

Chapter 20: To Infinity and Beyond 201
Chapter 21: Final Thoughts 205

Glossary of Historical Figures 207
Glossary of Terms..................................... 211
Bibliography... 219
About the Author..................................... 223

Chapter ONE

The Mysteries of the Universe

Oh, to wonder what is there.

DO YOU REMEMBER when you were old enough to understand the immensity of the universe, and you found yourself far away from the city lights, gazing into the infinite night's sky, and beheld the magnificence of that starry night with all the brilliant celestial bodies on display? What did you think about, or were you just silent with amazement? Do you remember where you were when you witnessed this spectacle, and who you were with? Did the vastness of the cosmos make you feel insignificant, or did the quiet beauty of the twinkling star light leave you with a sense of belonging to something wonderful? Do you remember wondering how all the heavens came into being, or if they had always been there? And, of course, we all wondered how really big were the seemingly infinite heavens? Remember and reflect on your experience, as I share my encounter with this awakening.

As a boy I remember going on summer camping trips with my family in California's Sierra Nevada, far beyond civilization and the city lights. My father would drive us in our old Pontiac, with a trailer in tow, deep into the mountains where he would precariously drive that last mile over a solid granite slab rock down into a meadow where our parents would set up a campsite under a grove of pine trees next to a large meadow and a gently flowing river. Very few people ventured to this meadow. Our campsite was 7,000 feet above sea level, which afforded us with clear and clean skies, comfortable summer days, the occasional afternoon thunderstorm, and perfect conditions for stargazing at night. For my sister Judy and I, this was heaven, something we looked forward to each year with excitement. These outings bonded our family together.

Our next-door neighbors back home, who had three boys, would all join us on these camping excursions. Together we camped far off the beaten path and we boys were given liberty from our parents to spend our days trout fishing from sunrise to sunset along a lazy river, only returning to drop off our trout and find something to eat. We also went on family hikes and horseback rides to backcountry lakes. Our group would spend some of our late afternoon's fishing at the beaver ponds, where we would occasionally see a beaver swim by, quietly patrolling his deep pond. The highlights of the day's adventures were told around the evening campfires.

We four boys slept under the stars at night in our sleeping bags without the need for a tent. When there was no moon, we watched the quiet cool brilliance of stars twinkling in the night sky. (See Illustration #1.) It was always a wonderment to behold and especially grand when a large shooting star would cross the night sky. I remember waking at all hours of the night, between dreams of fishing, to notice how the earth's rotation had changed the location of the stars. On warm, moonless nights we would all take our sleeping bags out onto the meadow and peer into the heavens from horizon to horizon. We were there during the 1960's when there were few commercial overflights at night and visible satellites were non-existent. No man-made influences to hamper our perspective. It was so quiet. We marveled at the Milky Way galaxy, pointing out the various constellations, and puzzled over how the universe had been formed. The experience of those nights left me with more questions than we had answers, and it left me with a thirst to want to find answers to these questions.

As we grew, many of these mysteries were explained to us. However, others still remained, such as: How big is the universe? Is it endless, or does it stay confined to a given space? What formed the universe or was it always there? And lastly, is there a connection between science's understanding of the universe and the biblical creation? The answers to these questions have evolved and changed over the centuries. How are we to know that the current understanding of our universe is correct?

So how do we develop the skills we need to find the truth about the cosmos? As young children, most of us began our consciousness surrounded by our parent's world, within the comfort of their loving arms. But as we grow out of childhood, our minds expand their understanding of a larger environment. At first, we learn about our family, our community, and then, through our formal education, we learn about things we haven't ever seen or touched. We learn, not only about current events, but also about historic information that helped shape our world into what it is today. During this time, we begin processing our learned knowledge to help us formulate our own understanding of the universe. We find it natural to ask questions to help in this process and draw our own conclusions.

For most people, the understanding of their universe reaches certain boundaries or plateaus, most likely set by environmental and cultural limitations, while others accept philosophies that guide their beliefs. Let's face it—with the challenges of daily life, most people are too busy to become experts on this subject matter or take the time to analyze and formulate their own creation vs. evolution hypotheses. Most people accept what others have analyzed or what has been disseminated though religious teachings. In both cases, people tend to become closed to new ideas and other possibilities. For example, some people have held onto an outdated concept that the Earth is flat. They formed a society called "The Flat Earth Society" that, despite a daily barrage of satellite information and photographs from space, still believes the Earth is flat. This group seemed to focus on a philosophy that if you can't personally view something yourself, then you can't accept new and changing information. Others have a difficult time accepting the possibility of an infinite cosmos, as the concept of infinity may be difficult for them to comprehend. It therefore makes it more palatable for most people to accept the current thinking that our universe was created in the Big Bang, which theorizes the formation of a finite universe. Our current-day scientific community is focused on how this Big Bang could have started rather than questioning whether this theory is still valid.

There are also those in our society who, by education and desire, have become labeled as scientists and philosophers studying the universe. It's one thing to be educated in a certain field, but quite another to truly have the capability to analyze and decipher the mysteries of the universe. It seems like many of these people take pleasure in developing wild and creative solutions that may tickle the mind or sell a few books along the way, but most of these theories have little factual evidence to support them. One of the more popular concepts has come from Carl Sagan who stated, "The universe is all that was, all that there is or ever will be." Of course, his beliefs, which also theorized that there was no Creator or creation, were derived from a biased opinion of all the facts surrounding the formation of the universe.

Our review of history will show that there have been some who developed a thirst for expanding their understanding of the complexity of everything around them and have focused on how this might fit into solving the mysteries of the universe. These analytical individuals, when given new information about objects within the cosmos, attempted to interpolate how this information might change the current understanding about the complexity of our universe. They have relied on facts instead of feelings. These innovative thinkers were able to solve complex problems by learning as much as possible about the subject they were studying. They then analyzed and visualized how all components within the problem work together. When they identified parts of the technical puzzle that didn't fit together, they challenged assumptions used in creating the existing models and attempted to formulate possible alternative solutions. Once these alternative solutions were identified, they were tested to identify the best workable solution.

Solving the mysteries of the universe can be accomplished in the same manner. In fact, I think some of the recent discoveries have shown that the current model of our universe has several flaws which deserve to be examined to see if there is a more accurate solution.

This book will attempt to analyze the current theories about the creation of the universe using factual information about the subject and then derive the most probable conclusions.

To start our journey, I feel it's important for us to review and dissect the historical record about humanity's theories of the universe and note how these interpretations have evolved. We'll see if any patterns develop from this review that will lead us to believe we should reassess some of our current accepted assumptions.

We will then review the mechanics that drive everything within the universe to see how that might support the different creation theories. This will be followed by a review of the relevant theories on creation of the universe.

Finally, we will challenge each of the creation theories by testing them against any and all relevant facts about the universe, some of which are fairly recent. By testing these theories, we should be able to eliminate inaccuracies, which should lead us to a more accurate understanding of the universe.

Let's start our journey!

CHAPTER
TWO

THE EVOLUTION OF THE GREAT CREATION

To arrive at the proper destination, one should make sure they've received accurate directions.

DURING MY THIRTY-SIX years working as a mechanical/nuclear engineer in the electrical generation industry, I spent my apprenticeship as a supervisor of power plant engineering at various generation facilities. It was a very rewarding assignment that taught me a great deal about the details of that industry. The engineers who worked at the various power plants were often called upon to figure out why a system or piece of machinery failed. If we were able to identify and remedy what caused a component failure, we could design a permanent fix or solution and hopefully avoid similar failures in the future. To assess the causes for some of these more complex failures, we would employ a technique called "root cause analysis" that forced us to re-evaluate all the assumed causes for the failure and, through dissecting and analyzing the information and assessing their validity, we were able to solve and remedy most problems. Later in my career when I moved into higher positions, I still found it useful to employ these same techniques to reach the best possible outcome on many an issue.

We will utilize these same techniques throughout this book to analyze the various theories behind the creation of the universe to see what we can learn. This will mean we will need to challenge everything we know about these theories, including the assumptions that have formed the basis of our current theory about the creation of the universe.

Before we begin this analysis of the different theories about how the universe began, it is important for us to first focus on the historic evolution of the universe from man's perspective. After all, these concepts drove our early beliefs and if some of these concepts were factually flawed, then they may have steered us in the wrong direction. We must also admit in this review that within our psyche, different stories or theories about the universe may have influenced our perspective. So we will need to make sure we rely on factual evidence to steer us in the right direction.

The postulation about how the universe or cosmos was either created or evolved has been with us throughout mankind's recorded history. Most of us can recount several of these hypotheses. These historical accounts have been plagued with inaccurate or incomplete information because of a limited observational perspective. As time passed, mankind invented new devices that allowed a more detailed view of the universe. These new perspectives help to dispel inaccuracies and thus allow for a more accurate interpretation of the universe. Surprisingly enough, these new interpretations often only solve a portion of the great mystery or at times create new misinformation, which is handed down for generations until it also is corrected by more advanced observations and analysis.

In this chapter, we will review and analyze each of the relevant historical theories to see what we can learn from their shortcomings. We will also attempt to correct, through analysis, some of the latest disinformation and attempt to formulate a more logical and complete picture. Let's take a look at the track record.

Geocentric Universe

In the sixteenth century BC, Mesopotamian cosmology believed in a flat and circular Earth enclosed in a cosmic ocean. In the sixth century BC, the Babylonian world map showed the Earth also being flat and surrounded by a cosmic ocean. They also believed that the land mass of Earth was swimming on the water and overarched by a solid vault of firmament to which the stars were fastened [1]. These early theories about the universe focused on a geocentric belief that the Earth was the center of the universe and that the sun, moon, and stars revolved around it. There was no distinction between stars and planets as they were assumed to be one and the same, only noting that some stars moved over time. Did this geocentric belief indicate a desire by early man to seek ultimate significance within his world, or was he just unable to conceive other options? Certainly, he would have felt that his feet were planted on solid and unmoving ground, so why wouldn't other objects move relative to his perspective?

---

[1] "Timeline of Cosmological Theories", Wikipedia, https://en.m.wikipedia.org/wiki/Timeline_of_cosmological_theories

These early beliefs also postulated that there was a large ocean that was beyond the sun, moon, and stars. There is some logic to how this great ocean in the heavens could have been theorized. What did these early observers see when they traveled to the ocean? They saw a vast expanse of blue. These waters also seemed very deep to them, as the ocean bottom dropped off into unknown depths not far from shore. Was there even a bottom to them?

To that same observer, what did they see when they looked into the sky every day? They saw a blue sky. And when the sun set and all light was taken away, the sky, like the oceans, turned dark. They also saw the moon and some bright stars during the day when the sky was blue and could have concluded that the blue was behind all the stars. In other words, they thought the heavenly waters were beyond the celestial bodies and were illuminated by the sun.

They also witnessed rain falling from the sky. Did they think that the water fell from the blue ocean in the sky to the ground below? They would not have noticed that rain clouds were formed by the evaporation of water from the Earth's surface. They would not have understood how raindrops form. They just assumed they fell from the heavens.

These early observers had no idea how far away the blue sky was from the Earth or the celestial bodies. So the reference to "waters of the deep" in the book of Genesis could mean over all the cosmos. After all, early man had no concept about the limits of our atmosphere and the vastness of space beyond. It is interesting, but fully understandable, that these early perspectives of the universe focused on large objects in our solar system and didn't offer any speculation about the immensity of the cosmos. In fact, since there was a belief that the heavens beyond the sun, moon, and stars were a deep ocean, they may have assumed that nothing existed beyond that point.

It isn't until the fifth century BC that Anaxagoras postulated that the moon was a sphere illuminated by the sun. Why did it take early man so long to understand that the moon was a sphere? If these early philosophers had observed the shadowing effect candlelight makes on a spherical object, they could have easily concluded the moon was spherical and lit by the sun. We can conduct this same experiment with a soccer ball and a flashlight. Anaxagoras also theorized that the sun and the stars were fiery rocks and that the moon was earthlike.

In the fourth century BC, Aristotle theorized that the Earth was also a sphere. Did he come to this conclusion based on Anaxagoras's work or was this deduced by his own observations that all heavenly bodies were round in appearance and most likely spherical? He may also have noticed, as Columbus did before he sailed in AD 1492, that sailing ships disappeared over the horizon as they sailed further away from land. It wasn't until 240 BC that Aristotle's theory was proven.

The Genesis Account

One of the earliest and most noted writings about creation is from the Bible in the book of Genesis. In Chapter 1 verses 1-2, the author describes the creation of the universe as follows: *"In the beginning God created the heavens and the earth. And the earth was formless and void, and darkness was over the surface of the deep; and the Spirit of God was moving over the surface of the waters.*" In this Genesis account, Moses goes on to declare that God also caused the stars and planets to take shape and created all the life forms to come into existence.

Since these words were put to pen, mankind has tried to understand exactly how all of creation or evolution of the universe could have occurred. Mankind has also attempted to discern throughout history whether there is evidence that points to a divine hand in the creation of the cosmos.

But what can we surmise from these Genesis writings? When it states "*In the beginning God created the heavens and the earth. And the earth was formless and void*", it implies that God created all the matter that is spread throughout the universe, but in such a fashion that there was no form to it, and therefore there initially weren't planets or suns for light. This seems to imply that all the atomic particles were instantaneously created and dispersed, but not yet combined into shapes, as it describes this matter as formless, or so small that it had no form.

Genesis also notes that after this beginning, "*God's Spirit moved over the waters of the deep.*" The term "moved" would leave the impression that God was watching over and shepherding His creation into the celestial bodies we see today. There is no mention of time involved. But if one is to move or hover as noted in other translations, that would seem to imply this occurred over a period of time, which could have been longer than many scholars originally imagined. Or it could be said that God set in motion the formation of the stars, planets, galaxies, and Earth over a longer period of time in order to achieve His purpose. We'll discuss in a later chapter the next set of verses in Genesis that deal with the six days of creation (or what's currently debated as "New Earth" vs. "Old Earth"). But for now, note that the first two verses may not be tied to the first calendar day of creation.

How do we interpret what Moses was conveying when he wrote in Genesis 1:2 *"and darkness was over the surface of the deep; and the Spirit of God was moving over the surface of the waters"*? Could the terms "deep" and "waters" mean over all the universe, or is it referring to a physical layer of water suspended above Earth's atmosphere? We'll debate this issue in a subsequent chapter. However, if earlier observers (as discussed above) felt that the blue-sky universe was water in a great ocean, did Moses also believe that when he wrote the book of Genesis? The Bible wouldn't be in error if he used this reference, as he was only defining it in terms that would be understood by peoples of that time.

After all, early man had no concept about the limits of our atmosphere and the vastness of space beyond. How else would God describe the creation to a man of that time? Would He first describe quantum physics, the vastness of the universe, the role nuclear fusion has in creating light, or would He describe it in terms that could be understood? He probably decided to keep it simple and in terms understandable to people of this time. After all, I assume His goal was to simply establish that He created all of existence, and not the mechanics of how He created it.

The first chapter of Genesis does not infer where the Earth was placed within the universe. It focused primarily on Earth's creation, the coming of man, and his relationship with God. This is a good reminder for us on where our focus should be as well. We may now discover in this book that we have sufficient evidence to understand how the cosmos was formed. The practical question is this: If we solve the mysteries of the universe, and that solution points to the Genesis God as the Creator, do we have an even greater responsibility to help our fellow man and to love the Lord our God with all our heart, soul, and mind?

Heliocentric Universe

In the third century BC, Aristarchus of Samos placed the sun as being the center of the universe with the Earth orbiting it. This is better known as a Heliocentric solar system within the universe [2].

Eratosthenes of Cyrene, who was the head of the Great Library of Alexandria around 240 BC, was the first to determine that the world was round and was able to accurately calculate its diameter [3]. This was accomplished by measuring the length of a shadow cast on the ground by two identical vertical rods at the same time but separating them by hundreds of miles on a north-south trajectory. This determined the curvature of the Earth and thus its circumference. Eratosthenes also calculated the distances between the earth, moon, and sun. Eratosthenes's discoveries became known to Roman society and were expanded upon by Claudius Ptolemy [4].

---

[2] "Timeline of Cosmological Theories", Wikipedia, https://en.m.wikipedia.org/wiki/Timeline_of_cosmological_theories

[3] "Timeline of Cosmological Theories", Wikipedia, https://en.m.wikipedia.org/wiki/Timeline_of_cosmological_theories

[4] "Timeline of Cosmological Theories", Wikipedia, https://en.m.wikipedia.org/wiki/Timeline_of_cosmological_theories

The Renaissance Period

During the Renaissance, Greek texts were rediscovered and among them were Eratosthenes' and Ptolemy's deductions [5]. Yet to openly state the Earth to be round was generally perceived as a contradiction to the Bible and anyone who postulated the Earth to be round became a subject for The Inquisition. So why did the church adopt a geocentric perspective and pronounce heresy upon anyone who opposed their view? Certainly, Genesis provides no scientific basis for a geocentric solar system. It was the explorations led by Columbus to the West Indies (American) in 1492 and concluding with Magellan's circumnavigation in 1522 that finally convinced most of the world and the Church that the world was indeed round [6].

With this mystery solved, greater attention could be paid to the heavens. At this same time, with the development of telescopes, advances were made in understanding the cosmos. In 1543 Nicolaus Copernicus, known as the father of astronomy, theorized that the Earth was not the center of the cosmos [7]. So this Heliocentric perspective was back in play.

Copernicus also identified the inner and outer planets, from Mercury to Saturn. But this was not the first time these planets were identified. They were also identified by the Babylonian astronomers

---

[5] "Timeline of Discovery of Solar System Planets & Their Moons", Wikipedia, https://en.m.wikipedia.org/wiki/Timeline_of_discovery_of_Solar_System_planets_and_their_moons

[6] "Timeline of Cosmological Theories", Wikipedia, https://en.m.wikipedia.org/wiki/Timeline_of_cosmological_theories

[7] "Timeline of Cosmological Theories", Wikipedia, https://en.m.wikipedia.org/wiki/Timeline_of_cosmological_theories

in the second century BC as well as by Aristarchus of Samos [8]. This dispelled the notion that these bodies were moving stars.

Not everyone was convinced the world was round. The Flat Earth Society, founded in 1956, promotes the false belief that the Earth is flat. I remember that following the Apollo moon landings in 1969, someone sent their society photographs of the round Earth from the moon. There were skeptics within this society that declared the photos were doctored. This is evidence of being blinded by their own feelings instead of analyzing the facts to see if their opinion should change.

Moving the Edge of the Universe

Early observers placed the edge of the universe at the edge of the visible night sky. Aristotle theorized about there being a finite universe, but an infinite timeline. Others disagreed. This was debated over the centuries until telescopes could show an even larger universe.

There was an ancient philosopher who theorized that no matter how large you declared the universe to be, you could also perceive it to extend further. This postulation assumed that the dimension of space extends forever, without end, or what we would call an infinite cosmos. Giordano Bruno, an Italian philosopher in the

---

[8] "Timeline of Discovery of Solar System Planets & Their Moons", Wikipedia, https://en.m.wikipedia.org/wiki/Timeline_of_discovery_of_Solar_System_planets_and_their_moons

late 1500's, believed in an infinite universe with infinite matter and numerous inhabited worlds [9].

Today, with modern telescopes, astronomers can see what appears to be the edge of the universe, but is it? Why wouldn't space continue on forever and, if so, why would the visible universe be confined to such a small portion of infinite space? We simply may not have the ability to see the extent of the cosmos. Think about that... There may be no end to matter and space!

Some have argued that if space and matter extend forever, then the night sky would be bright as there would be a sun located in every speck of the sky. This is absurd because sunlight from distant suns would have such a faint signature that it could not be seen here on Earth. A recent confirmation of this occurred when they trained the Hubble telescope on what looked like a dark segment in the universe. When they exposed this darkened area in space to the capture of light for a prolonged period of time, they discovered approximately 10,000 distant and young galaxies, some of them thirty billion light years away, meaning they were at least thirty billion years old. (See Illustration #2.)

Others have postulated that the universe wraps around itself or is shaped like a soccer ball. So if you traveled far enough in one direction you would eventually come back to your point of origin, but from the opposite direction. Are these philosophers unable to perceive that space and matter may have no end?

Due to the popularity of the Big Bang theory, people have assumed that a finite universe exists. We'll discuss this further in

---

[9] "Timeline of Discovery of Solar System Planets & Their Moons", Wikipedia, https://en.m.wikipedia.org/wiki/Timeline_of_discovery_of_Solar_System_planets_and_their_moons

another chapter, but for now just imagine how small our known universe would be in the infinite dimension of space. Also, why would our universe be the only object in that infinite space?

Darwinism

Charles Darwin put a real twist into creation theories with his proposals about how life evolved on this planet. His teachings popularized a hypothesis that the world and universe had evolved and were very old. But how old were they?

It's a big stretch to have observed what appeared to be individual species adaptation and then to conclude that all life evolved. For instance, just observe all the intertwined and complex systems in your body without even mentioning the complexities of DNA and cell division. It's extremely unlikely that natural selection through mutation would evolve such complex organisms. Darwin also admitted in his writings that if biological science were more complex than their primitive understanding at the time, his evolutionary theories wouldn't hold water.

It's an even bigger stretch to conclude that the universe evolved based on observations about plant and animal life. But to his credit, Darwin did discover evidence that the world appeared to be much older than believed. He noted that volcanic activity may have been involved in the creation of Earth's continents, and he also felt that erosion appeared to take place over a long period of time.

Galaxies are Discovered

With the advent of telescopes, astronomers were seeing further into the universe. In 1755 AD, Immanuel Kant asserted that what scientists had originally believed to be nebulae were really galaxies, separate from and independent of the Milky Way Galaxy. William Herschel, thirty years later, proposed the theory that our sun is at or near the center of the galaxy [10]. Nice try, William!

In the 1860's, William Huggins developed astronomical spectroscopy, which determined material composition through analyzing the spectrum of light given off from distant objects. He went on to determine that the Orion nebula is mostly made of gas, while the Andromeda nebula (later called the Andromeda Galaxy) is probably dominated by stars [11]. (See Illustration #3.) As a high schooler in physics class, I remember debating whether the atomic structure of matter was, in fact, universal across the universe. Huggins's observations demonstrate that these distant objects in the universe are made up of identical particles to what we find on Earth. His work has helped to prove that all matter is built from the same building blocks.

The Great Mathematicians and Physicists

In the last 130 years, there have been many wonderfully gifted people who have diligently tried to define the universe and everything in it. These scientists would debate various subjects,

---

[10] "Timeline of Cosmological Theories", Wikipedia,
https://en.m.wikipedia.org/wiki/Timeline_of_cosmological_theories

[11] "Timeline of Cosmological Theories", Wikipedia,
https://en.m.wikipedia.org/wiki/Timeline_of_cosmological_theories

postulate new theories, and then try and prove their guesstimates by expanding the laws of physics through the derivation of mathematical equations. They would hope that someone in the future would find sufficient evidence that would prove their theories. One of the most famous and renowned explorers in this field was Albert Einstein. Einstein believed he could derive a mathematical linkage between all the known physical sciences, like matter, energy, gravity, and time. Here are a few of his postulations:

- **Matter and Energy Linkage**

    One of Einstein's most famous equation is $E=mc^2$, where energy (E) and mass (m) are linked together by the speed of light (c) squared as a multiplier. Loosely translated, this states that you can get a whole lot of energy from the conversion of a little bit of matter. Einstein believed that matter and energy were interchangeable. He set off on an exhaustive mathematical derivation to prove his point. He felt that matter could be converted to energy and that this process could also be reversed to convert energy into matter. His observations about the sustaining energy release on our sun could have led him to believe this linkage possible.

    This revelation helped unearth the development of the nuclear age, as scientists believed there would be great potential for energy creation if they could harness the energy contained within matter, which was a process they believed was occurring on the sun's surface.

    More evidence about the validity of this theory has been unearthed since Einstein developed this Theory of

Relativity in 1915, and I'm surprised this theory hasn't been widely disclaimed. We'll discuss this theory about matter and energy linkage in another chapter.

- **Speed of Light**

Einstein also theorized that as you approached the speed of light, something would prevent you from exceeding that speed. He also felt that the mass of anything approaching the speed of light would increase, and time itself would slow down. This phenomenon somehow prevented light photons from traveling faster than the speed of light. To accept this theory, he felt that space somehow has a reference point or invisible force that prevents travel faster than the speed of light. So, in other words, no warp speed Scotty!

Is this really the case? Is there really some invisible force that can tell an object in the vacuum of space (where there is no reference point) to slow down or not exceed the speed of light? Or do light photons have a common property that, when they're generated and leave an object (like the filament in a light bulb), they exit the filament at the speed of light? Which is more probable? We'll address these questions about Einstein's theory in another chapter. But I would like to make a point about this and other theories from Einstein. They are all theories with no physical evidence to support them. They only exist as part of an extensive derivation of mathematical equations.

- **Light Photons Are Particles**

    Einstein was convinced that light photons were small particles and not simply rays of light or electromagnetic waves. Einstein also believed that these small particles were not impacted by gravity, but merely traveled in a straight line through space, unabated, forever. This was later partially proven through the discovery in space of what is called an "Einstein Ring" (See Illustration #4), which is the visible distortion of light when it passes a large object in space. This proved that light waves were also particles, as Einstein theorized. It also proved that these same particles of light were affected by gravitational fields in space which could change their direction and even possibly slow them down. This will be discussed later as well.

- **Gravity Wells**

    Einstein theorized that all objects in space were located inside what he called "gravity wells", which attracted matter and kept the planets in stable orbits around the sun. Einstein believed this, even though he had no evidence to prove his theory. We'll discuss the validity of this hypothesis in another chapter.

- **Universe Always Existed**

    Einstein believed that the universe has always existed. This has been debated back and forth. The Big Bang theory has brought this into question. We'll discuss this in a different chapter after we've had a chance to examine how we should determine the most probable solution.

Quantum Physics

Quantum physics or quantum mechanics has been postulated and advanced since the early 1900s. It attempts to define the behavior of matter and energy on atomic and subatomic particles. It's believed to explain how everything works. This science attempts to explain particle interactions through the derivation of mathematical models. It's an interesting science but is it factual and does it play a role in the creation of the celestial bodies we find in space? It seems more realistic that the simple but well-defined role of atomic particles alone accounts for the formation of everything we observe in the universe.

The significant role that gravity plays in the coalescence of matter cannot be underestimated as well. Given the design of everything that can be detected and defined in the universe, we can see that with enough time, these building blocks could have formed the universe we see around us today. We don't need to theorize a new science to find an abstract and complicated solution.

Large Telescopes Reach Out to Greater Distances

In 1610, Galileo brought the Milky Way and near-Earth objects into a more precise perspective by building a telescope that could magnify images in the night sky tenfold. It was, however, the large telescopes built in the twentieth century that really brought the known universe into perspective. Astronomers were able to see much deeper into space and found countless galaxies in the universe.

In the 1920s, Edwin Hubble, using the telescope at the Mount Wilson Observatory, was one of many astronomers who were mapping out this new frontier of discovery. In 1925, he noticed that particles of light from distant galaxies were "redshifted" [12]. Redshift meant that the electromagnetic wavelength or the frequency of the light was longer or stretched out, which shifted the incoming light waves to the red spectrum of light. (See Illustration #5.) Therefore, these light particles were approaching Earth at a velocity slower than the speed of light.

After observing this phenomenon on multiple objects in space, he determined that the farther objects were from the Earth, the slower their light particles became when detected on Earth. In trying to explain this finding he had two options. The first option would be to assume that all of the galaxies were rushing away from one another, which would indicate the universe was expanding. This would eventually lead to a Big Bang theory about the creation to the universe. The second option assumed that light photons, which are particles of light, slowed down ever so slightly as they traveled through space.

---

[12] "The Fabric of the Cosmos: What is Space?", NOVA, Season 38, Episode 16

Why did Hubble choose the first option? It appears he did so after being influenced by Einstein's theory that light photons travel throughout space at the speed of light and were not affected by gravity or other forces. It's interesting to note the profound differences these two options represent. However, I have not discovered any research reassessing this decision, even after it has been proven that light is impacted by other forces in space.

Early theorists held a fixation of their placement in the center of their known universe, only to be proven wrong. Where do you suppose the Milky Way Galaxy is located within today's ever-expanding universe model? Astronomers and scientists say the universe is about ninety-three billion light years across [13] and has no center. Really? Are they afraid to admit that in their Big Bang model the Milky Way galaxy and therefore Earth is near the center of the known universe? Their deep space observations in all directions show the edge of what they can see at equal distances from Earth, so in other words, we are at or near the center of the universe. How convenient! If you're interested, there are several 3D mapping simulations of the known universe that can be found on the internet. Note their placement of our galaxy. We'll discuss this in another chapter.

Black Holes Discovered at Galactic Centers

In the last several decades, astronomers discovered what appeared to be dark spots in space where there was no light being emitted. They assumed that these locations contained objects that

---

[13] "The Physics of the Universe - Where in the Universe is the Earth?",
https://www.physicsoftheuniverse.com/where-in-the-universe-is-the-earth.html

were made up of extremely dense material. This material was so dense that nearby objects, and even light photons, were unable to escape their gravitation pull. They are referred to as black holes. (See Illustration #6.)

Recently, they have discovered that galaxies contain several massive black holes at their centers. These giant black holes would create a massive gravitational pull on their galaxy. This explains why rotating galaxies reacquire solar systems that are flung out into their spiral bands.

So what makes up these black holes and how are they formed? Will they eventually suck the entire universe into darkness? We'll discuss this in another chapter.

Small Particles Have Mass

I remember taking nuclear physics in college. At that time, in order to balance out the equation for nuclear reactions, neutrinos were inserted into the equation. They were believed to be so small they could pass through miles of solid rock before they would collide with an atomic nucleus. These elusive particles were believed to have no mass and only contain the physical properties of momentum and spin. Like early studies of light photons, physicists eventually were able to discover that these particles exist and have a very small mass component as well.

Currently, scientists believe there are some forty-five subatomic particles at play in our universe, though many of these are only theoretical. Are they real, and how can their existence be verified? Finally, do these tiny particles play any role in the formation of the universe other than to shoot about the cosmos and balance out physics equations?

Dark Matter/Energy

Dark matter is postulated to be a nonluminous material that exists in space and could take several forms, including weak-interacting subatomic particles or high-energy randomly moving particles [14]. It is theorized that these particles were created right after the Big Bang and help explain why there are areas of the universe that don't appear to have any visible matter. They also believe that dark energy is pushing the universe apart, helping to expand large voids in our universe that separate concentrations or bands of galaxies. The assumption is that eighty-five percent of all matter in the universe is made up of these tiny particles [15]. That's a big guesstimate with no evidence to back it up. There are other explanations for the appearance of matter distribution and as to why there are voids in space that we'll discuss in another chapter.

String Theory

String theory is one of several theories that attempts to define parallel or multi-universe possibilities, but there is not a thread of evidence to prove their existence. Let's stick with what we can identify and let the dreamers and sci-fi novelists keep on dreaming. We might find out that we have a never-ending cosmos

---

[14] "Dark Matter", Wikipedia, https://en.m.wikipedia.org/wiki/Dark_matter
[15] Nola Taylor Redd, "What is Dark Energy?", https://www.space.com/20929-dark-energy.html

with infinite space and matter, which means we wouldn't need to invent ideas for sub-sectioning the cosmos into lesser entities.

Multiverse Theory

The multiverse or Many Universes theory describes a possibility that there may be additional universes created by their individual Big Bang beginnings. In fact, some theorize there could even be an infinite number of universes in an infinite space. There is no evidence that this is a reality, but it is interesting to note that scientists are beginning to look beyond our universe into the infinite cosmos. If the Big Bang expansion theory were proven to be false, and we continue to find matter even further into outer space, one could also conclude that there could be infinite matter in an infinite cosmos that is only driven by gravitational forces.

Higgs Field Analogy

In 2012, Peter Higgs theorized that there are subatomic particles or a field of energy throughout space that causes some atomic particles to slow down. There is no evidence to back up this theory, but it is interesting that there are those trying to explain why matter slows down in space. Have they looked at gravitational effects?

Planets Discovered in Other Solar Systems

Scientists have recently discovered planets surrounding other suns. This area of discovery is expanding rapidly. What puzzled me about the initial discovery of these planets wasn't that they found them, but that they were surprised other planets outside our solar system exist around other suns. Do people believe that our solar system, or even the elements that make up our solar system, are unique within the universe? They shouldn't.

Gravitational Waves

For several years I've wondered what the speed of gravity might be. For example, if a new object were suddenly placed in space, how fast would the gravitational force from that object reach out and begin pulling other distant objects. The gravitational force wouldn't instantaneously reach across the universe, would it?

Gravity is a force emitted by matter in the form of a gravitational field. We don't see it, and to us, we don't see a change in its effect or force unless the mass of an object changes, which doesn't occur around here unless you've fallen off your diet. But how fast does gravity reach out and is there a maximum acceleration for gravity? If gravity can only pull an object at a fixed velocity, say close to the speed of light, then gravity could have a dampening effect on light photons traveling through the universe. We'll discuss this further in another chapter.

Gravity waves have been theorized for the last one hundred years. It was felt that these waves would be created by the sudden and massive density changes from the collapse or collision of large objects in space. Within the last two decades, advancements in gravity wave detection devices have not only proven their existence but also provided details about their velocity and location of origin. We'll discuss the relevance of three recent findings in another chapter, but for now note that these finding prove that gravity waves travel at the speed of light, unabated by other influences.

Observations Derived from This Chapter

So what can we conclude from reviewing these historical discoveries and concepts about the universe? What do we see as patterns, and what do we see about our ability to formulate an accurate understanding about the cosmos? Let's summarize these observations below:

- **Mankind Is Naturally Inquisitive**

    Humans will question anything they don't fully understand and seek to find answers. The success of mankind has largely been driven by this ability to question the norms and seek out new solutions that better their lives. I'm sure early man focused more time on trying to survive and spent little time wondering about the cosmos. As societies evolved, they could expand the education of their individuals and support those who could expand the knowledge of that society. In the Dark Ages, when mankind was again more focused on survival, the Catholic Church had a bigger influence on setting norms. The Church seemed to be hindering the expansion of knowledge until global exploration stretched their view of Earth's position in the universe.

During the Industrial Revolution of the late nineteenth and early twentieth centuries, great minds focused on invention and innovation. This led to devices that could see further into the heavens than ever before. Brilliant scientists, mathematicians, and philosophers were able to exponentially expand their interpretation of this larger cosmos and even attempt to define things where no evidence previously existed.

- **Perspectives Were Sight Focused**

Although our history is full of mythical theories, the understanding of our universe has mainly been driven by visual observation. Early theorists focused more on what they saw with the naked eye, so their universe appeared to them as very tiny. An exception to this was the debate surrounding an infinite universe, which still rages today. As telescopes and other devices peered further into space, our universe expanded in size, and the edge of it seems to be at the edge of what can be seen. Could this universe expand further than what we see today?

- **Theories Abound**

I've tried to include in this chapter some of the key historical findings and theories about our universe. It would take the rest of my life to include in this book what I've uncovered in my research about all the different theories that relate to the cosmos. Our review has demonstrated that many of the historical theories have been debunked or proven to be in error. So what other theories about the cosmos will be proven false or misleading?

- **Assumptions Are Rarely Reevaluated**

    People develop theories about the universe after viewing evidence, but the assumptions behind these theories are rarely reevaluated once the theories are accepted. Scientists accept them as fact and then build on them with their next theory. Theories are not reevaluated when new evidence comes to light. We shouldn't assume someone else is correct in their prior assumptions. Several examples have been discussed in this chapter and will be elaborated on later in this book.

- **Where Do We Fit?**

    There has been a tug-of-war regarding where we fit in this universe. Are we at the center or not? Historically, man has typically envisioned solutions that place him at the center of his universe. First, everything revolved around the Earth. As we gained further perspective, we found ourselves part of a larger universe in which we became increasingly insignificant specks in a large ocean of space. But even now with the Big Bang theory we find ourselves back at the center, although scientists won't admit it for fear that this centrist theory will be debunked as a repeat of failed theories about our significance in the cosmos.

- **The Search for Significance**

    Another human trait is our search for significance. Each of us have different desires for the legacies we leave. Personally, the older I get, the more I desire to be remembered. Some of us try to leave legacies for our children and grandchildren. Others look for significance in their professions. I'm sure scientists would like to be known for finding a cure for cancer or for discovering some revelation about the cosmos. However, our individual searches need to be genuine and fact-based and not based on some far-fetched speculation. For example, imagine someone seeing a soccer ball in his backyard and then theorizing the universe must be shaped like a soccer ball rapping in on itself because he likes the shape and wants to be remembered for some fantastic revelation!

- **Historical Bias**

    Have historical perspectives and assumptions about the universe clouded our ability to search for accurate outcomes? If some of these historical views are found to be in error, we could be led in the wrong direction when we seek to expand on their theories.

- **Speculations Abound**

    Speculative stock buyers get criticized for purchasing stocks on a hunch or feeling that a particular company will bring them a fantastic reward. Too often they lose their shirts. Their speculative theory is, "Got a hunch, bet a bunch." One becomes more successful in the stock market when they do a lot of research on a company before they purchase any stock. Recently, there have been many speculators about the universe. I would use this phrase to describe them: "Got a hunch, write a bunch." Most of these speculators seem to be trying to sell books and make a name for themselves.

- **Contracting or Expanding Universe?**

It appears that this debate has continued for some time, but it's been more acceptable by scientists and astronomers to focus on expansion following the acceptance of Hubble's expanding universe assumption. But what is really happening? We will explore the options.

- **Hindsight Is 20/20**

From reviewing this chapter, you can see that perspectives of historical events are always better than predictions about current or future events. We'll need to keep this in mind as well.

CHAPTER
THREE

How Do We Know?

Before you solve a problem, you need to understand the question.

ALL OF US remember growing up in grade school where we found ourselves at times having to solve written problems. We quickly learned that the key to finding the solution was making sure we understood the problem well enough to translate it into a workable equation. The rest was math. Similarly, we need to make sure we don't jump to a conclusion without thoroughly analyzing all the evidence supporting each option. After all, we want the correct solution, not the most popular or tantalizing one.

Also, remember our discussion in the previous chapter about root cause analysis, which challenges and analyzes all data and assumptions to derive the correct solution to complex problems. I believe we can utilize these tools to help establish some key evaluation criteria to use in our analysis about the universe.

Our brief historical review has shown us that there have been ever-evolving concepts about our universe and that most of these concepts have been in error. How do we know if our current understanding of our universe is correct? How should we then proceed to evaluate the different current projections about the universe and determine the most probable solution to this mystery? I believe we should first distill down the lessons learned from past mistakes and utilize these to help us set up effective tools or guidelines that will assist us in finding the most probable solution. When doing so, we find three distinct areas we can utilize to help us in our review. They are as follows:

1. **Things Aren't Always as They Seem**

    With evolving technology that brought visual evidence into clarity, many of the historical assumptions were thrown out or revised. Physical evidence has proven to be the best way of finding the truth. We need to first determine if there is physical or visual evidence before we draw conclusions. Then our observations need to be correctly interpreted. It is important to weigh all facts and evidence before we conclude our findings relative to visual sightings.

2. **The Simplest Answer Might Be the Best Answer**

    Absent of sufficient physical evidence, we should look at what logically makes sense and eliminate improbable and complex ideas using Occam's Razor. Wikipedia states that Occam's Razor, often expressed as the law of economy or law of succinctness, is a principle that generally recommends selecting the competing hypothesis that makes the fewest new assumptions when the hypotheses are equal in other respects [16].

For instance, all possible options we consider must be sufficiently explained by available evidence in the first place. Of the surviving options that meet this test, the remaining option with the fewest new assumptions would most likely be the correct solution. We don't need to be grandstanding with some complicated model that has impossible odds of being realistic. Therefore, we will invoke Occam's Razor as one of our test methods.

3. **Challenge Assumptions**

Assumptions used to support each option need to be assessed for their accuracy. We have seen several examples where past interpretations of assumptions have led to additional errors down the line, if not corrected.

Secondly, do all the assumptions we develop, or use fit together and support any new models of the universe? Keep in mind that even though you may have a strong conviction about your assumptions, they might be incorrect.

---

[16] "Philosophical Razor", Wikipedia, https://www.google.com/search?client=firefox-b-1-d&q=Wikipedia+%E2%80%93+Philosophical+Razor

Chapter FOUR

The Elephant in the Room

*Your brilliance has no bearing on reality when
your feelings get in the way of facts.*

I REMEMBER IN my college physics class debating and questioning the validity of linking totally separate and distinct equations about different variables (like time, gravity, the speed of light, mass, and energy) when the professor walked us though how Einstein had formulated his famous equations behind his Special Theory of Relativity. We soon found out, through a lengthy discussion, that no one should question Einstein's brilliance. If we had exposed any fault in his analysis, it would be like exposing the emperor as "having no clothes". So we moved on, despite the elephant in the room.

Over the years I found myself reassessing the debate about several of Einstein's findings and feel it's time to revisit this debate. After all, if we fail to challenge the validity of past assumptions, we may find ourselves trying to explain away factual evidence that contradicts an originally flawed theory.

Einstein desired to discover a universal nexus or equation that linked everything together within the physical world. His Theory of Relativity, which was published around 1915, fell short of this goal but was hailed as a major breakthrough in advancing the theories of physical sciences. Some of his most famous theories linked matter and energy, as well as properties associated with traveling at the speed of light. Einstein, however, was never able to achieve his lifetime goal of finding a linkage between everything. This does raise an interesting point. Even though Einstein desired to find this linkage, which would have been the ultimate accomplishment for a mathematician, it doesn't mean that one exists. So why should everything be linked together? Do matter and energy need to be interchangeable?

During Einstein's time, he and other mathematicians and physicists were pondering theories about the universe and physical sciences, and then attempting to prove their theories through the derivation of mathematical models. They would then look for physical evidence in an attempt to validate their hypotheses. In the early 1900s, physical sciences were in their infancy, which limited the available evidence to not only prove their theories but also may have directed some of these theories in the wrong direction. Back then the speed of sound hadn't been broken as airplane travel was relatively new and, of course, subsonic. Nuclear physics was accentually nonexistent as no one quite understood the fusion interactions that power our sun, and Hubble had not yet discovered the slowing of light photons from deep space.

In my working profession, we found ourselves at times attempting to derive a solution to a complex problem though economic analysis. You had to be extremely careful in your analysis to use the most accurate assumptions, as well as the most pertinent model for outlining the problem. We found that if you weren't careful, you could distort the reality of the problem and make an unwise decision. We would jest about some of the unwise or distorted projections by saying, "Liars figure, and therefore figures lie." The real challenge was to sort out fact from fantasy. These early pioneers attempting to solve theoretical questions over a century ago also needed to make sure they weren't playing a shell and pea game when attempting or seeking to prove themselves correct.

We'll look at three critical theories of Einstein in this chapter and weigh them against more current physical evidence to see if we need to update their conclusions.

Matter and Energy Linkage

As mentioned earlier, one of Einstein's most famous equation is $E=mc^2$, where energy (E) and mass (m) are linked together by the speed of light (c) as a multiplier. This conclusion was derived through a complex derivation of equations. Einstein theorized in this equation that matter and energy are linked and interchangeable. He believed that matter could be converted to energy and energy could be converted to matter.

If we investigate what may have influenced Einstein to reach this conclusion, we find that he believed the cosmos had always existed and that there must therefore be a way for matter and energy to be interchangeable so that the nuclear reaction that illuminates our sun could last forever. If there wasn't this ability to convert matter and energy back and forth, then the cosmos would eventually burn out and consolidate.

Let's look at how energy is released in a nuclear reaction. When studying nuclear physics in college in the 1970's the professor explained that the combined particles within an atomic nucleus appear to have less mass than the sum of the sub-components that make up that same atom. This difference is called the mass defect. They also believed that this mass defect (Δm) must have occurred from the conversion of binding energy (ΔE) into matter. They believed they could utilize a derivative of Einstein's equation $\Delta E=\Delta mc^2$ to calculate how much binding energy would be needed to break apart an atomic nucleus. This separation could only be done when sufficient binding energy was available to bombard a given nucleus. There are some informative videos on the internet that help to explain this process [17]. It should also be noted that this modified interpretation of Einstein's equation left intact the basis of his finding but fundamentally altered the intent of Einstein's derivation, which was to show the complete and reversable conversion of mater to energy and energy to matter.

To help explain the relevancy of binding energy look at binding energy divided by the number of nucleons as graphed and shown on Illustration #7. This ratio is not constant and varies for different elements. The graph shows this ratio as compared to the number of nucleons per atom or the atomic mass of an atom. When atoms with low atomic masses on the left side of the graph are fused the result will yield an element with a higher binding energy per nucleon and thus be able to release excess energy. An example of this is the fusion reaction of hydrogen we see on

---

[17] "For the Love of Physics – Nuclear Binding Energy", https://www.youtube.com/watch?v=BYRz_9wvJzA

our sun's surface. As hydrogen is fused into heavier elements it will eventually reach an atomic mass where the fusion reaction will not be sustainable because it will require as much or more energy to fuse the heavier elements as the reaction would generate. However, this does not preclude the ability to fuse heavier elements which could be occurring within our sun as there is an abundance of energy available to do so.

For atoms on the right side of the graph, like uranium, it's fission process would result in two smaller elements with higher binding energy numbers, thus an increase in energy release like what is seen in nuclear reactors. The byproducts from this fission reaction would eventually decay and become stable elements.

Let's summarize some facts about this relationship between matter and energy as we currently understand nuclear physics.

- We can say that energy is released by the fusing or splitting of atomic nuclei.

- We know it takes a certain amount of binding energy to separate or combine particles within an atomic nucleus. To provide this binding energy it was theorized that protons and neutrons would need to convert a portion of their mass to energy. What didn't make sense at the time was the notion that these particles could be changing their mass.

- We also know that the energy release we see from matter doesn't come from the complete elimination of atomic particles. The heat generated by our sun in its massive fusion reaction cycle does not destroy atomic particles.

- We know that the fusion and fission reactions in the universe are driven by the attempt of matter to reach a neutral state while interacting with other matter in the gravity driven process of combining matter into larger objects. As an example, our sun's hydrogen will fuse to heavier elements which will eventually transform our sun into a black hole.

- We also know that the amount of energy released from the fusion or fission of different elements is not proportional to a given quantity of matter as theorized in Einstein's famous equation. For example, the energy release from the fusion of hydrogen is far greater than the release of energy from and equivalent mass of plutonium in a fission reaction.

You may ask what difference does this make? Well, this is key to determining if the universe has always been in existence or has a finite life. Einstein was correct in theorizing there was great potential for energy as it relates to matter, like in a nuclear reaction. But it wasn't how he envisioned it. There is no reversable cycle of energy and mass conversion.

If by some miracle you could convert energy to matter, what matter could you produce? Would it be electrons, photons, or neutrons? How would that energy transformation know which component of matter to produce and how would you assemble enough energy to make this transformation take place? It doesn't make any sense.

A recent article [18] outlined what is believed to be three basic mistakes in Einstein's interpretation of $E=mc^2$. The basics of the first finding is that particles within the nucleus of an atom have the combined properties of being both kinetic energy and mass. When an atomic particle is accelerated to a specific velocity the kinetic energy inherent to that velocity contributes to the overall mass of the particle. This also holds true in the reverse when a particle decelerates. I find this very compelling.

The second finding in this article notes that small particles contained within the atom which are released during nuclear reactions have a small measurement of mass contrary to earlier beliefs. Their small mass had not been included in determining the mass defect of an element.

These first two findings in this article make it very clear that in nuclear reactions there is no partial or complete elimination or creation of atomic particles. These findings provide evidence that Einstein's Special Theory of Relativity is no longer relative. If you are interested in understanding more about these findings and how they disprove Einstein's theory I suggest you review this material.

Since Einstein's famous equation has a false premise that matter must be eliminated to create energy, and it does not consider that there can be different amounts of energy released from the nuclear interaction of different elements, his Special Theory

---

[18] James Carter, "The Living Universe - A New Theory for the Creation of Matter in the Universe", https://living-universe.com/questions-and-answers/how-einstein-was-wrong/

of Relativity should be replaced by stating that when you split or combine certain atoms you cause a nonlinear energy release.

Speed of Light

Einstein also theorized that as you approached the speed of light, time slowed down and the mass of an object increased. This somehow prevented light photons from accelerating faster than the speed of light. To accept this theory, he also felt that space somehow had the ability to detect and prevent anything from traveling faster than the speed of light. However, is this really the case? Is there really some invisible force out there that can tell an object in the vacuum of space, where there is no known reference point, to not exceed the speed of light? Or do light photons have a common property that when they're generated and leave an object, like the filament in a light bulb, they simply exit the filament at the speed of light? Which is more probable?

Let's also look at the mindset that may have been behind this theory. When Einstein was a young man, he observed that sound waves could not exceed the speed of sound. This was true even if an approaching object generated sound waves. But if an object moved away from you, the sound waves would reach you at a reduced speed, which would be less than the speed of sound. For example, consider the sound we hear from a train whistle. As the train approaches us, we hear the pitch of the train whistle at the same tone as if the train were standing still. We therefore conclude that sound waves are prohibited from exceeding the speed of sound in our atmosphere. But when the train passes us, the sound waves from the whistle change in pitch to a lower tone, noting that the sound waves are traveling at a slower speed.

Einstein became convinced that there was something preventing sound waves from accelerating faster than the speed of sound, but also noted that sound waves could travel at slower velocities. Little did he know that this phenomenon was the result of particle interaction. Sound doesn't actually travel in waves, but instead is accelerated by particles colliding in sequence, thus pushing the sound from air molecule to air molecule radiating out from the sound source. The denser the medium you attempt to send sound waves or shock waves through, the higher the traversing velocity. This is why shock waves travel faster through water or the Earth than they do through the air.

During Einstein's time, no one understood what would happen if you traveled faster than sound waves in our atmosphere. We've all heard the stories that people believed you couldn't travel faster than the speed of sound in our atmosphere. So it may have been theorized that some force was preventing sound waves or anything else from going faster than the speed of sound.

Also, Einstein may have felt that light traveled in a similar fashion as sound waves and that these two phenomena should exhibit similar responses to their environments. I believe Einstein was influenced by his observation of sound waves, which led him to predict and try to prove a similar result for light waves traveling through space.

The big difference between sound waves and that of light photons traveling at the speed of light is their interaction, or lack thereof, with other particles. We know what limits sound transfer, which is driven by particle interaction to affect the transfer of sound. But light photons are particles that move independently of anything else. Also, since space is a total vacuum, there are no other particles to interact with except for the force of gravity.

Under Einstein's theory, if you were traveling in a spaceship at the speed of light relative to earth and shined your flashlight in the same direction you were traveling, the light beam would never leave the front of your flashlight. So how would the beam of light know it was already traveling at the speed of light, because it's only perspective within the spaceship would be that of not moving?

Let's take Einstein's speed of light theory one step further. If the universe is expanding and objects within that universe are moving apart and generating light photons at the speed of light at their source, what would we see? We would observe these photons approaching us at different, but reduced, velocities according to Hubble. Einstein predicted, there is an invisible force in space that limits acceleration to the speed of light. How can this force measure and adjust itself across an expanding universe unless it is only acting as an initial dampener? There couldn't be an invisible force spread out across the cosmos that limits light photon acceleration if you are constantly having to adjust the reference point where these photons were generated in and expanding universe.

The Hubble telescope has helped to resolve this mystery. Cosmologists have recently discovered the existence of over one thousand objects in space traveling faster than the speed of light. This evidence proves there is no limiting velocity in the cosmos and that Einstein's equations were in error, most likely due to flawed assumptions about observed particle behavior.

With these equations debunked about light travel we could also conclude that there is no increase in the mass of particles or that time would slow down as an object's velocity approached the speed of light. These equations for mass and time, as they relate to the speed of light, were all linked together. So if one equation is found in error it's most likely the rest of these hypotheses will be dismissed. I understand that they believe to have proof that time slows down as your velocity increases, but I doubt this will hold up under Einstein's equations.

This finding that eliminates the restrictions on light travel also calls into question whether Einstein's famous equation about Special Relativity is still valid, as its derivation was influenced by Einstein's flawed perspective about light acceleration. So, $E=mc^2$ may no longer be valid.

You can derive all kinds of equations to try to support a theory that was heavily influenced by your environment and limited knowledge at the time about the science you are studying. But as new facts are discovered about what influenced your theory, you need to reexamine your assumptions.

To summarize, there is no evidence that any force restricts movement past the speed of light and there is now factual evidence that matter does move faster than the speed of light. It also appears that light photons and other accelerated particles begin their journeys at the speed of light. So, "Captain Kirk to Scotty, give me warp speed!"

Gravity Wells

Einstein was puzzled as to why the planets appeared to be in stable orbits around the sun. He wondered why their orbits didn't decay and eventually be drawn into the sun. Were objects in space located where they are because they are inside a gravity well, and are these wells keeping the planets from being sucked into the sun? Einstein believed so and theorized that all objects in space are located in gravity wells.

When Einstein developed this theory, little was known about the orientation of celestial bodies and whether these objects had been located in stable orbits. At that time, the Milky Way galaxy was the only known galaxy in the universe. He certainly didn't know about the existence of black holes and their ability to hold galaxies together. He also believed that the cosmos and everything in it had always existed, and therefore there needed to be a mechanism that prevented everything from coalescing in space. Since then, scientists and astronomers have mapped the universe and have discovered all sorts of different celestial bodies at play, as well as the evidence that gravitational forces are working to eventually coalesce and collapse the universe.

We also know that all matter includes a component of gravity which draws other objects together, and we know that the sun's gravity similarly tries to pull objects toward it. To counter these forces, the planets maintain their separation due to centrifugal forces. Remember that planetary objects are in frictionless stable orbits, so there is nothing in space that would slow and decay their orbits, which would draw them into the sun or eventually eject them into space. We have also monitored our moon's orbit and have noticed is has slightly changed over time.

So there does not appear to be any eternally stable orbits held in place by gravity wells out there maintaining a constant separation of celestial bodies. The only forces at play are gravitational attraction and centrifugal forces. But I guess you could also label the gravitational field that objects in space create as being within their own gravity well. It may just be a matter of semantics.

Conclusions

So what do we conclude from this discussion? Obviously, Einstein was a brilliant man, trying to solve complex problems though mathematical analysis. He was before his time in postulating about multiple areas within science and physics. One wonders what he would have theorized or concluded had he been born 100 years later.

What I find interesting is the lack of factual evidence, even now, to support any of his theories. I would also argue that, as time has passed, we see more evidence that contradicts his findings. If we were to apply Occam's Razor to determine if these three postulations were valid, we would conclude, by evidence alone, that Einstein's theories would not survive. Therefore, we need to now question if these theories have adversely influenced other postulations about the cosmos to make sure we find a realistic solution about the infinite horizon.

Chapter FIVE

What's the Matter?

To understand the infinite, you must first understand the finite.

To help solve the riddle about the formation of the cosmos, it is important for us to sort out if matter changes over time and whether it's universal throughout the cosmos. This could provide evidence about whether the cosmos has a beginning and an end, or whether it has always been there. Let's take a look at four critical observations about matter.

Size of Atoms

I remember back in high school someone asking the chemistry teacher how he knew atoms were the smallest particles. Could the particles in the atom be made up of even smaller objects? The teacher responded that atoms were so tiny there couldn't be anything smaller. There was no debate. The scientific community believed they were the smallest, therefore we should accept their conclusion. But what wasn't discussed was whether all of these components that make up atoms were individually the identical size throughout the universe.

That actually brings up a good question. Are all of these atomic particles individually the same size? For example, are all neutrons identical in size, and are all electrons much smaller but of identical size? How can all of the matter in the universe be identical in size and configuration or design? The teacher could have responded with a better answer than, "It's because they're the smallest particles." We know on earth that all matter is made up of atoms and that their subcomponents are of separate but identical size.

Looking at distant stars in the outermost reaches of the universe, they have detected hydrogen and other elements from spectral analysis of star light, which provides proof that the building blocks for all of matter across the universe is identical. On a larger scale, we can observe similarly configured galaxies throughout space, which supports this same theory that all matter is identical. We now know that, as far as we can see into the universe, all matter is of the same size and made up of the same elements. So we can conclude that all matter in the universe is identical in size and function.

The current understanding from the scientific community has accepted this same principal. The current theories surrounding the Big Bang creation theory make this easier to accept, as they believe all matter was identical and began its existence from a singularity in space and therefore could be of an identical design. But what if the Big Bang were proven to be wrong, and all matter was brought into existence already spread out throughout the universe? Especially if it was an infinite universe with infinite matter? Why would it be of the same size and function, unless its origin came from an infinite Designer? Remember, it's not just the smallest matter in the universe, it's the identical design, function, and size of matter throughout the cosmos.

Structure of Atoms

The structure and function of atoms are amazing and miraculous! It's the perfect design. By combining three independent and indestructible atomic particles, you can form totally different elements. Not even the heat from a burning sun or a supernova can disintegrate these atomic particles. Atomic fusion or fission reactions are the only way to change atoms from one element to another. Oh, yes, you can also bombard certain elements to accept an atomic particle into its nucleus, but when doing so they, more often than not, become unstable and may eventually decay back to their original configuration.

One of the properties of the atomic structure is the release of energy that lights our sun and warms our planet through a sustaining fusion reaction that converts atoms into heavier elements. These reactions occur on large bodies in space, which are dense enough to begin a fusion reaction.

On Earth, we use heavy elements like uranium and plutonium to generate thermal energy in nuclear reactors. These heavy elements are extracted through mining and refining operations, but these same elements are also some of the heaviest elements on our planet and would also have migrated during the Earth's formation to its central core. Earth's core could be laden with uranium, which would help to generate thermal energy, keeping the central parts of this planet's core molten. This is an essential building block for enabling a rotating magnetic field core within our planet, which protects the Earth's atmosphere from being carried off by solar winds.

Electrons orbiting atomic nuclei maintain a balance of positive and negative charges within stable atoms. Some elements can easily transfer electrons to generate static electricity or provide us with the ability to transfer mechanical energy into electrical energy, which drives our society's ability to electrify our homes and industries. Just think about the conveniences you would lose if you didn't have electricity in your home, or even the ability to generate a spark to ignite a fire on a cold night.

Atoms, besides containing positive and negative charges, also contain a gravitational force. Without gravity, none of the matter in space would combine. Space would be occupied by a cold dark cloud of diffused particles.

Matter and Energy Conversion

Throughout the universe we see a display of nuclear fusion that sends out photons of light from distant stars. So if fusion of hydrogen creates the release of energy, what will happen when all the hydrogen and lighter gasses are fused together? Will there be a reversal? As far as we know, there is no reversal to this fusion process. The current theory supported by visual evidence is that at the end of a sun's life it will either explode in a supernova and the remnant will become a white dwarf star or a black hole. Smaller suns will transition to neutron stars. (See Illustrations #8-13.) The aftermath of each will leave these remnants as being much smaller with denser cores. It's also believed that material ejected by any explosion will fuse atoms into even heavier materials. This would support the notion that we will see an overall depletion of the lighter gases needed to sustain fusion reactions. So there is no everlasting, never-ending conversion of matter to energy and then energy to matter over and over again.

So what is the endgame? Will this fusion process go on forever, or does this provide evidence that the universe will eventually burn out and go cold and dark with only the continuing existence of massive black holes? Scientists recently agreed that the universe will eventually burn out. But wait a minute, if there is an ending to the universe, then there had to be a beginning! And if so, was it a creation like the Big Bang theory, or was it an instantaneous creation of matter spread throughout an infinite cosmos?

There is another theory floating around that a reoccurrence of the Big Bang could be happening. This theory proposes that after the universe burns out, gravity will pull all matter together and eventually create a reoccurring Big Bang that continues on forever. It's interesting to note that no one has any idea how this reoccurring Big Bang could occur.

We'll discuss this further in other chapters, but for now we can conclude that matter has NOT been dispersed throughout the universe for all eternity, and that there is an end to the generation of energy from the nuclear reaction of particles.

The Composition of Objects in Space

What is the composition of objects in space, and how are they formed? During the last several decades there have been debates within the scientific community about the composition of objects in space. Recent evidence has revealed that all of these objects in space are comprised of atoms, so we don't need to elaborate on that any further. We will discuss the existence of dark matter and what role, if any, it plays in the composition of objects in space in another chapter.

So how are objects in space formed? It's kind of like a cascading avalanche of combining smaller objects into bigger, more dense ones through a process of gravitational attraction. First of all, the universe could have started its existence as diffuse dust particles, which would have been drawn into clouds. This would have been observed as a Genesis beginning, where there was no form to the creation. Gravitational forces would have eventually caused the dust clouds to combine into solid objects, like some of the dust and rock asteroids we see today in our solar system. These dust clouds would have been made up of all of the different elements.

Could this beginning have started from a universe filled with only diffuse hydrogen gas? Hydrogen gas in invisible and would also meet the definition in Genesis of a universe without form. In this scenario, it could have been drawn together by gravity to form dense clouds in an extremely cold space. However, it's ignition to form a sun in a sustainable nuclear reaction could not occur as it would not be dense enough. As it turns out, suns need a dense core made of solid material with a strong enough gravitational pulling force to keep their fusion reaction closely contained, and thus sustainable.

We've also observed vast areas of dust clouds in our Milky Way galaxy. It's believed that these same areas are stellar nurseries that gather enough material to form suns as well as capture, through gravitational forces, enough material to form planets within young solar systems.

As more and more solar systems are created, gravitational forces start drawing them together to form clusters of stars, and eventually contain large numbers of stars to be classified as globular clusters. (See Illustration #14.) When suns burn out or go nova, they create denser objects that attract even more stars. When globular clusters are attracted to one another, their collision will start a rotation that will form a small galaxy. As additional galaxies collide, more massive galaxy will be formed. (See Illustrations #15-17.)

Within these galactic formations, the collapse of stars will eventually combine and fuse into black holes. They have recently found that galaxies contain several massive black holes at their center, which helps to explain why matter within the grip of a galaxy never escapes its gravitational pull. And of course, over billions of years these galaxies, after sucking in their neighboring galaxies, will eventually have no other debris to draw into their massive black holes. This will transform these galaxies into massive black holes and give the appearance of emptiness, since no light or energy will escape their grasp.

Chapter Six

The Gravity of the Situation

Before you reach for the stars in your analysis, it's important to understand the forces at play in the universe.

LET'S TAKE A look at gravity from a microscopic perspective. One of the remarkable characteristics of atoms is that they have a component force of gravity. This force of gravity draws particles together. Each particle within the nucleus of an atom has an equal amount of gravitational pull. The heavier the element, the larger its gravitational force. As far as we know, there is no boundary on how far this force reaches out to attract other material. Even though the strength of a gravitational field doesn't change, its force effect on other objects is exponentially reduced the further removed it is from those objects.

We can see gravitational forces at play in a spiral galaxy by observing that all matter within that galaxy is being first flung out from its center, only to be pulled back into the confines of the galaxy by gravity. (See Illustration #15.) We have also noted that galaxies are being pulled toward one another through gravitational attraction. Our Milky Way galaxy and its neighboring Andromeda galaxy are being pulled together by gravity over vast distances and will collide some five billion years from now.

So I think it's safe to assume that gravity is a force that has no boundary to its apparent reach. When applying this to all the matter in the universe, we find gravity to be the primary force that determines the movement of everything.

It should also be noted that the force of gravity is a somewhat disconnected force. Meaning, it can only loosely attract other objects. It's more akin to the force of wind blowing past an object, where it's only able to affect movement of that object due to the frictional bond between the object and the surrounding air particles. In other words, this gravitational force can only slowly accelerate objects that come under its attractive influence.

But how fast do gravitational forces from an object reach out and begin pulling distant objects toward it? Does this force extend out from an object at a very high velocity and then have the ability to pull and accelerate other objects toward it until they match the limiting speed of gravity from the attracting object? Or do objects instantaneously send out their gravitational forces across the cosmos and have the ability to accelerate distant objects toward them with no limit on their ability to accelerate these objects?

The ultimate goal in this discussion will focus on trying to resolve these questions on how fast gravitational forces can accelerate objects traversing the universe. Because if gravitational forces can only pull something so fast before it has no remaining effect on that object, then there is a limit to gravitational acceleration.

Speed of Gravity

The velocity or speed of gravity has been hard to detect and prove until recently. The term "gravity wave" is used to describe a gravitational wave that is generated and spreads out from its source by a sudden change in the density of a large object in space. It's kind of like a wave ripple you see when you throw a rock on a pond, except that a gravity wave radiates out in a three-dimensional sphere. For the last 114 years, scientists believed theoretically that gravity waves existed and that they could provide evidence as to the speed of gravity. This could be initiated when massive objects in space collide, like black holes, as their combined mass would suddenly change due to a massive and sudden energy release.

Recently, they have devised an accurate way of measuring these gravity waves that travel through space with the installation of two Advanced LIGO facilities in the United States. These high-tech facilities have measured the velocity of these gravitational waves and have determined that they travel at the speed of light. They've also measured the rapid change in mass caused by these cosmic collisions and have noticed a massive energy release, which includes a significant release of gamma radiation. It's interesting to note that they have not seen any visible light emanating from these events. Could it be that light photons take longer to travel to Earth, or does the gravitational pull from these events preclude the ejection of light photons? To begin to answer these questions, let's focus on three recent gravity wave events to see what we can surmise.

- On September 14, 2015, a gravitational wave was detected having traveled some 1.3 billion light years to Earth [19]. It was believed that this gravity wave was caused by the collision of two black holes, and that this gravity wave was traveling at the speed of light.

- On August 17, 2017, scientists were able to determine the speed of gravity when they recorded a gravitational wave that was caused by the collision of what they believed to be two neutron stars. This event occurred 130 million light years from Earth, and they noted that gamma radiation from that event was detected just 1.7 seconds after the gravitational wave arrived [20]. This discovery proves that the speed of gravity is identical to the speed of light, as gamma rays travel at the speed of light.

  What I find interesting is the fact that in their analysis of this neutron star collision event, they found that the gamma radiation from this event arrived slightly later than the gravity wave. They noted that one possibility for this delay could be explained by assuming that particles traveling in space could be affected by the gravitational fields they encounter along the way [21].

---

[19] "Gravitational Wave", Wikipedia, https://en.m.wikipedia.org/wiki/Gravitational_wave#History

[20] Ethan Siegel, "Ask Ethan: Why Do Gravitational Waves Travel Exactly At The Speed Of Light?", https://www.forbes.com/sites/startswithabang/2019/07/06/ask-ethan-why-do-gravitational-waves-travel-exactly-at-the-speed-of-light/?sh=3fedbbab32dc

[21] Ethan Siegel, "Ask Ethan: Why Did Light Arrive 1.7 Seconds After Gravitational Waves In The Neutron Star Merger?", https://www.forbes.com/sites/startswithabang/2017/10/28/ask-ethan-why-did-light-arrive-1-7-seconds-after-gravitational-waves-in-the-neutron-star-merger/?sh=9ab5c4775d46

They also noted that gravity waves would not encounter anything that could hamper or retard their velocity, which is why the gravitational wave may have arrived ahead of the gamma ray burst. This possibility would support the notion that particles traversing the universe are affected by gravitational forces.

- On January 5, 2017, another black hole collision was detected in the form of a gravity wave that had traveled three billion light years [22]. This represents a sighting that's nearly one quarter of the distance to the edge of the detected universe under the Big Bang theory. If this recent discovery also detected the gravitational wave as traveling at the speed of light when it reached Earth, then we have evidence that the universe is not expanding. Let me explain. In Hubble's studies about an expanding universe, he would have noted that the light photons traveling three billion light years would have arrived at a reduced velocity. Hubble concluded that these slower particles were evidence that the galaxies which generated them were moving away from us in an expanding universe. But if gravity waves, which are not altered by other influences, arrive here at the speed of light, then these

---

[22] Jonathan Amos, "Gravitational waves: Third detection of deep space warping", BBC Science Correspondent, https://www.bbc.com/news/science-environment-40120680

distant objects are not moving away from us and there is no expanding universe.

Scientists studying this new area of discovery are convinced that the speed of gravitational waves is not altered as they traverse the universe. If in fact the universe is expanding, some scientists also believe that gravity waves could see their relative velocity, as detected on Earth, decline in a reaction like the redshift effect light photons experience when they travel vast distances. Again, I find it interesting that there is no mention of any deceleration of the gravity wave discovered in the August 17, 2017 observation. In this new area of discovery, it will be interesting to see if gravity wave velocity mapping in our universe will show an expanding or coalescing universe.

What I find interesting in this debate about gravity waves is that despite the massive release of energy in these cosmic collisions, there is no visual evidence of their occurrence. You would think that if gamma rays were emitted from these events, you would also see the ejection of light photons in a bright flash occurring at the same time. Could it be that light photons are more susceptible to gravitational forces than other particles? If so, maybe they're still in transit and we'll eventually be able to visually witness these events.

Also, would light photons be affected differently by gravitational forces than gamma rays? Highly energetic gamma rays pack quite a destructive force, giving them the ability to penetrate deep into solid material. Conversely, light photons, with their extremely small mass, have almost no ability to penetrate nontransparent material and more commonly bounce off objects they only illuminate. So does gravity have more of an ability to affect light photons as they traverse the universe? It will be interesting to see if cosmologists follow up on observations pertaining to these neutron star and black hole collisions to see if and when they might detect light photons emitted from these events. This could help determine the effects of gravity on light photons. We'll discuss in a later chapter the effects of gravity on the kinetic energy of light photons. But, for now note that a prolonged gravitational exposure could elongate the electromagnetic wavelengths of light photons, giving them the appearance of moving at a slower velocity.

As a side note, I find it interesting that the speed of gravity has the same initial velocity as light photons, electrons, magnetic field forces, gamma rays, and a myriad of other particles. It's an interesting observation about the mechanics of particle acceleration, or maybe it's a design-imposed limit within the creation of everything. I would label this designed velocity "The Goldilocks Zone" velocity. After all, if these particles and forces traveled too fast or too slow this would lead to unstable or inept fusion reactions.

Gravitational Effects on Light Photons

Light has dual properties of being an electromagnetic wave as well as being a photon particle. In this discussion, we'll focus on the effects that gravity has on the velocity of light photon particles.

For gravity to reach out and begin accelerating an object in its direction, it must first reach out at the speed of gravity. Once this force reaches an object, it begins to accelerate this distant object in the direction of the gravitational source. As gravitational forces begin to attract this object, the force of gravity would have an identical velocity component, but this time, instead of reaching out, it's now pulling the object. Think of it this way—if gravitational forces reach out at a certain velocity to begin accelerating objects toward it, it also makes sense that it would pull and accelerate that object with an identical velocity. These velocities would be equal but opposite, so that the force of gravity would be in equilibrium. Therefore, the speed of gravity is equal when extending out and when attracting objects.

As an example, think of a juggler who is constantly throwing balls in the air. The balls leave the juggler's hands at the same speed as they return. In other words, a balanced reaction. A visible example of a balanced flow of forces would be to observe the magnetic forces at play around our sun. You can observe these arcing bands of magnetic fields carrying particles that reach out and come back to the sun at incredible speeds, which are comparable to the speed of light, but they are coming and going at the same velocity, thus demonstrating the balanced flow of forces. Are gravitational and magnetic forces that dissimilar in their balance of flows?

Can we then assume that gravitational forces have limits on their ability to accelerate objects? This could occur as different objects approach each other at a closing velocity that matches the speed of gravity. At these speeds, the gravitational forces would diminish their ability to further accelerate these objects because gravitational forces have an associated maximum velocity component that only has the ability to accelerate objects that are traveling at a slower closing velocity than the speed of gravity.

Think of it this way—when you stick your arm out the window of your car you feel the force of the wind push your arm backward. This would be similar to objects at rest being influenced by the full acceleration force of gravity. But if the wind is matching the speed of your car and is going in the same direction, your extended arm would not feel any force being placed on it. In a similar way, gravitational attraction forces have almost no ability to influence the acceleration of objects that are matching their speed of gravity.

So how would gravitational forces impact the velocity of particles traveling at the speed of gravity? I think it would be safe to assume that particles, like light photons, traveling toward an object would be minimally attracted by that object's gravitational pull.

With this conclusion, we should also look at the effect of gravitational forces on objects moving away from each other. In this case, as objects move away from each other, they would experience the full force and effect of each object's gravitational field or force, as they are pulling in opposite directions and the speed of gravity would not dilute the force's net effect. This doesn't necessarily mean that the force applied to each object would be any greater than the actual gravitational force, it just means that there would not be a dilution of the force's effect.

We should note, however, that if gravitational forces can be applied to the physics equation $F=ma$, where force (F) equals mass (m) times acceleration (a), we could assume that there may be a similar relationship with gravitational forces being proportional to the mass of an object times the acceleration of gravity. This would lead us to the conclusion that there would be an increase in gravitational forces on objects pulling away from each other because their separating speed of gravitational forces would be greater than the speed of gravity. There would also be a decrease in gravitational forces when objects are being drawn toward one another because their closing velocity would be less than the speed of gravity.

This explains why objects traversing the universe slow down. It's not that there is a barrier in space preventing objects, like light photons, from exceeding the speed of light, but it's gravity that is retarding objects from exceeding the speed of gravity.

Concluding there is no magic force or barrier at the speed of light in space that is restricting matter or photon particles from being accelerated past the speed of light, we can have Captain Kirk tell Scotty, "Give me warp speed, Scotty!" But when traveling through space, we would feel gravity's effects slow us down.

We also now know that gravity impacts light photon particles. Einstein and Hubble believed that photons of light were not affected by gravity and were instead held to a constant velocity equal to the speed of light. However, the later discovery of Einstein Rings, which illustrated the bending of light around large objects in space, also demonstrated that photons of light are affected by gravity. (See Illustration # 4.) So if gravitational forces cause light photons to change direction, can this same force cause their deceleration over time? If not, why not? A force is a force, of course, of course! If gravitational forces can change the direction of light photons, they can also interact and change their velocity.

Let's look at how this new theory about gravitational forces would work if you were a photon of light traveling at the speed of light across the universe. When you are traveling through space, you would have forces behind you and in front of you. If you start out traveling at the speed of light, the effect of gravitational forces in front of you, pulling you forward, would be minimal, as the speed of the gravitational forces would match your speed and have no ability to pull you forward. But the gravitational pull of objects you just passed would be able to apply a greater gravitational force, as they would be pulling in the opposite direction of your travel. The net effect would slow you down over time.

You could argue that as you traveled across the universe, there would be an equal amount of matter in front of you as there was behind you, which would cancel out their gravitational effect. However, that could only be true if you were traveling at a low velocity.

Conclusions

The evidence in this chapter shows us that recent evidence about gravity waves may prove that distant objects are not moving away from us. The evidence presented in this chapter points to gravity having a deceleration effect on light photons. So going back to Hubble's options, we would now conclude that he may have chosen the wrong option. The mapping of gravity waves in the universe should provide sufficient evidence to resolve this debate about an expanding universe. Also, the discussion in this chapter about the effects of gravity on light photons may make it look like the universe is expanding when it really isn't. We just don't know how much deceleration of photons occurs over time.

A physicist I know commented about the ability of gravity to slow down light photons. He noted that gravity was such a weak force in the universe that it could not possibly slow down light photons. In response, I would say, "Yes, gravity is a small force with minimal effect, but like the formation of diamonds, which are made by slowly crushing carbon over millions of years, the slow deceleration of photons could also occur after traveling across the universe for billions and billions of years."

The evidence that has been relied on the most to predict the formation of the universe has come from observations of visible light photons and from non-visible particles beyond the ultraviolet and infrared spectrums. Visual observations have proven to be the most accurate tool in helping us understand our universe. However, our review has shown us that mankind has historically erred in their interpretation of what it is we are observing.

Chapter Seven

How Really Big Was That Universe?

Can you comprehend an infinite cosmos?

THERE ARE TWO questions about the size of the cosmos that we will cover in this chapter. The first one will look at how expansive the three-dimensional realm of space really is. The second will address how much of that space is filled with matter.

Greek philosophers like Democritus offered the argument that if you set a limit to the size of the cosmos, one would only need to further extend your defined boundary to see that the cosmos is bigger than your original concept or imagination. This would lead you to the conclusion that you could extend the boundary of space forever and therefore conclude that space was infinite in size, as you would never find an end to the vastness of space. This revelation points to an obvious conclusion that should be given careful consideration. If you travel far out into the cosmos, even though you don't find other objects or if you come across a barrier, there is still the existence of space beyond those obstructions. Space has no end and is therefore infinite in size.

The size of space has been debated back and forth as we have seen in our review. It wasn't too long ago that the concept of an infinite cosmos was the accepted and (seemed to be) the most logical reality. But recently, the debate has refocused again on what they could visualize. The discussion about an infinite cosmos has become a background issue with the hypothesis surrounding the Big Bang theory. Attention is focused again on what can currently be observed within our cosmos. In doing so, cosmologists are wrestling with how such an event could have started, as well as trying to figure out if this finite Big Bang universe will spread out for eternity or collapse into a reoccurring Big Bang explosion. There has been little discussion on whether their assumptions used to develop their Big Bang theory need to be reexamined.

Many modern cosmologists also believe the universe is finite, largely driven from the fact that they can't detect any matter beyond thirty billion light years away. This observation is based on the discovery of cosmic microwave background radiation at the edge of our known universe. With the assumption that nothing else exists beyond this point, some cosmologists believe that one cannot see past the barrier of our detectable universe and that this must therefore be the edge of the universe. What is interesting is that in this concept the Earth is again at the center of this mapped cosmic microwave background.

Others believe that an infinite universe with infinite mass would be extremely bright. They are forgetting that light gets fainter the further it travels before reaching Earth. This was recently demonstrated when the Hubble telescope went looking at what they believed to be a segment of dark space. They found, after a long light exposure, that the area was full of young galaxies further away than previously imaginable. But also note, that if an infinite universe with infinite diverse matter were only thirty billion years old, then we would not see any evidence of material beyond thirty billion light years away, as the visual evidence of their existence would not have arrived here yet.

The Big Bang theory formulates that the universe began its existence 13.8 billion years ago. We now see galaxies on the edge of our universe, which are an equal number of light years away from Earth. For matter to now be visible from the edge of our universe, having first being flung out from the Big Bang, it would have to be much older than current forecasts. Viewing distant galaxies on the edge of our discovered universe would mean that it had taken an additional billions of years for them to first be formed, and we are now seeing these formed galaxies some 13.8 billion years later. Start adding up the number of years needed to support the Big Bang theory, and you'll find it is much older than the current projections.

So which do you think is more probable? Scientists have built a large enough telescope to see the edge or end of the universe, and nothing exists beyond that in an infinite but empty cosmos; or the universe is never-ending, and man just hasn't been able to see far enough? A third possibility would be that if an infinite cosmos and everything in it were created long ago, we would only be able to see a small portion of that universe, as light would not have reached Earth from objects in deep space. With man's track record of ever-changing and expanding discoveries about the size of the universe, the more likely reality is that we just haven't been able to see far enough.

There arose a dilemma after focusing so much attention on the Big Bang creation theory. Some of the scientists realized that if they focused only on matter being included in this one Big Bang creation, they were limiting their universe to a tiny fraction of space. If we tried to compare the size of the infinite cosmos with our finite universe, it would be like comparing the size of Earth to a grain of sand. The universe would be so tiny in all the cosmos it wouldn't be noticeable. Why have an infinite cosmos with only a finite universe? That would be a waste of space.

So out from the woodwork sprang numerous cosmic philosophers who theorize that the cosmos has different shapes, or that it could bend in on itself so it could fit within a defined space, or there could be multi-universes, or parallel existences, or worm holes to other realities. You name it. But wait a minute. Forget about past *Star Trek* episodes and get real. If we can see that the cosmos goes on forever, why is it so hard to comprehend that matter could also extend forever? What will really bake your noodle is the possibility that the infinite cosmos, with all the infinite matter scattered about within it, may have been created in an instant.

Chapter Eight

Dark Matter Is No Matter at All

*You can't get something from nothing, and
how can you know what you can't see?*

DARK MATTER WAS first theorized in the 1930s by the Swiss astronomer Fritz Zwicky, who felt that the mass of known matter in galaxies was not great enough to generate the gravitational forces to hold a cluster of galaxies together. He also noted that each independent galaxy moves at too great a speed for galaxies to remain in clusters. Yet these galaxies were not spinning away from each other. He felt that they had to be held together by gravitational fields created by an undetected mass[23].

More than forty years later, American astronomer Vera Rubin found that the same principle is true within a single galaxy. The mass of stars alone does not exert enough gravitational pull to hold the galaxy together. She also discovered that stars in the far

---

[23] "Dark Matter", Wikipedia, https://en.m.wikipedia.org/wiki/Dark_matter

reaches of our galaxy rotate about the Galactic Center at the same rate (revolutions per billion years) as stars close to the center. For our Milky Way galaxy, this rate is approximately four revolutions per billion years. Rubin concluded that there was some invisible, massive substance surrounding galaxies, exerting gravitational forces on all their stars [24].

In both of these theories, they felt there wasn't sufficient mass to hold galaxies together. This can now be explained away with the discovery of massive black holes located in the centers of galaxies. These massive black holes located in the centers of galaxies not only interact to contain rotating celestial matter within their individual galaxies, but also stabilize the rotation of stars within galaxies. These black holes are also strong enough to reach out and attract neighboring galaxies.

Furthermore, Zwicky's hypothesis about the gravitational pull between multiple galaxies can also be explained away when looking at the total gravitational pull within a cluster or line of galaxies whose gravitational forces would be in balance, thus keeping these galaxies in a semi-stable orientation. This is especially apparent when observing that these galaxy clusters are located within bands surrounding the massive voids in space. (See Illustration #18.) These clusters of galaxies could also be revolving around each other in elliptical orbits, thus preventing them from being drawn together.

Rubin's assumption that external gravitational forces influenced the rotation of galaxies has a fatal flaw. If there are external gravitational forces surrounding galaxies, these forces would be pulling the galaxies apart, not holding them together. Knowing

---

[24] "Dark Matter", Wikipedia, https://en.m.wikipedia.org/wiki/Dark_matter

the influence of black holes on galactic forces should have totally discredited the notion that dark matter exists.

Another dilemma arose when the cosmologists identified large voids in the universe. Their previous assumption about a young expanding universe being caused by the Big Bang would require that they find matter equally spread throughout the universe and drifting apart. They noted that if gravity were the driving force to create these voids, it would not meet their timeline for the Big Bang creation, as it would point to a much older universe.

They therefore went back to the assumption of dark matter and created dark energy as well, to help explain why these voids exist. They also theorized that dark energy might be used to postulate that an invisible energy force could be causing these voids to theoretically expand.

Of course, you would think that this force would be pulling objects together instead of pushing them away. They assumed that this invisible dark matter could be used to fill these voids within the universe, thus saving their Big Bang theory. But why would we have some regions of space with dark matter and others with galaxies. Do the two of them not mix well at the galactic dance party? It appears to me that trying to find the logic for inventing dark matter to fill these massive voids in the universe was eliminated once massive black holes were discovered.

To help you visualize these massive voids in space as well as the orientation of galaxies, I would suggest you look up the various Millennium Simulations on the internet that were published within the last sixteen years [25,26]. You'll find a wide variety of photos and videos that depict sections of our universe. This research attempts to illustrate how dark matter could have accelerated the matter in the universe into its current configuration. But you'll note that if you overlay the dark matter with the visible matter simulations, they are both in the same locations. So, it seems that there is a conflict when it comes to the distribution of dark matter in these simulations. We've learned in this chapter that cosmologists felt that dark matter was needed to fill the massive voids in space as well as influence galaxies. But in these studies, they assume the dark matter is in line with the galaxies, which leaves them with no explanation for how these massive voids could have been formed within the Big Bang timeline. So much for the supposition that the theoretical invisible dark matter would occupy the regions of the universe that appeared to be void of matter. Anyway, I do believe these visual simulations are very informative in illustrating the distribution of galaxies within the universe.

Interestingly enough, when they calculated the area in space being occupied by these voids, it represented seventy-five percent of all the universe. It's also interesting to me that the cosmologists have theorized that dark matter must represent roughly the same percentage of all the matter in space. I wonder how they came up with that number. Filling voids in space with the non-detectable dark matter to help you explain the flaws in the Big Bang

---

[25] "The Millennium Simulation Project MPA", Max-Planck-Institute, https://wwwmpa.mpa-garching.mpg.de/galform/virgo/millennium/

[26] Michael Boylan-Kolchin, Volker Springel, Simon D. M. White, Adrian Jenkins, Gerard Lemson, "Millennium Simulation-II", https://wwwmpa.mpa-garching.mpg.de/galform/millennium-II/

theory, instead of challenging the assumptions behind the Big Bang theory, has become an unnecessary distraction. You can't make this stuff up!

Also, there are some cosmologists assuming that dark matter is hiding evidence of our universe expanding beyond what we see. But if all of this dark matter exists, why isn't it impeding our view of distant galaxies? If dark matter's subatomic particles go zipping all around and interact with all known particles in space, why aren't light photons deflected by dark matter as they travel through space? If these interactions took place, we would see distant objects as blurred images.

It seems more logical that if the universe is older than the Big Bang beginning, but not infinitely old, gravitational forces alone would have pulled the diffuse matter into three-dimensional ribbons of galaxies, leaving wide areas void of any matter without the need for the Dark Side!

CHAPTER
NINE

Busting the Big Bang

Check your assumptions and emotions at the door when coming into an analysis.

THE BIG BANG creation theory was first proposed by a Belgian priest named Georges Lemaitre on January 18, 1917, when he theorized that the universe began from a single primordial atom. Lemaitre was a monsignor in the Catholic Church and had been fascinated by physics and studied Einstein's laws of gravity, published in 1915. Lemaitre desired to show a relationship between science and creation [27,28].

In the 1920s, Edwin Hubble observed that light photons traveling from distant stars slowed down by observing the redshift effect on light coming from distant stars. As he charted this phenomenon, he determined that the farther away these light

---

[27] "Georges Lemaitre: Father of the Big Bang", American Museum of Natural History (AMNH), https://www.amnh.org/learn-teach/curriculum-collections/cosmic-horizons-book/georges-lemaitre-big-bang

[28] "History of the Big Bang Theory", Wikipedia, https://en.wikipedia.org/wiki/History_of_the_Big_Bang_theory

photons traveled from in the universe, the slower their velocity reaching Earth. As stated previously, Hubble was faced with two possibilities to explain these sightings. The first was that, indeed, these photons of light slowed down the farther they traveled in space due to some resistant force. The second was to take an opposite perspective and assume that light photons were still leaving distant objects in space at the speed of light, but that these distant objects were accelerating away from Earth. Under this scenario, we would see their light photons arriving at a reduced velocity and assume that the universe was expanding and moving apart. He had a decision to make.

The scientific community at that time was adopting Einstein's theory that gravity would have no effect on light photons, as the universe itself limited the acceleration of objects to the speed of light. With this understanding, Hubble, not knowing about Lemaitre's theory, deduced that the universe was expanding. Lemaitre, hearing about Hubble's discovery, proposed again in 1927 that Hubble's evidence was, in fact, proof of a Big Bang creation from a single point in space [29].

This Big Bang creation theory has been adopted as fact ever since, with little review of the assumptions behind it. For scientists, they have been busy trying to figure out how this Big Bang creation could have started. They are also trying to find out if they can prove it to be part of an endless cycle of Big Bang creations, thus fostering their belief that the universe had no beginning and will have no end. The religious community has looked at the Big Bang creation as proof of God's one and only creation. These two

---

[29] "NASA – What Is the Big Bang?, NASA Science for Kids" NASA Space Place, https://spaceplace.nasa.gov/big-bang/en/

camps of science vs. religion are in a tug-of-war to prove if the Big Bang is never-ending or is truly evidence of God's one and only creation.

But remember, the one and only thread of evidence supporting a Big Bang depends on light photons traveling unabated across the universe. Despite all the scientific wrangling attempting to define how such an event could have occurred, the scientists have run into a few insurmountable problems. In this chapter we'll go through each of these issues.

1. Big Bang Beginning

To help resolve the validity of the Big Bang theory, we must first rewind time itself to see how such an event could have begun. Scientists believe it started with a tiny sphere of extremely dense matter that suddenly decided to explode outwardly. From a scientific standpoint, they are unable to figure out how such an explosion could have initiated. How does such a dense concentration of matter that is held tightly together by gravity suddenly disregard gravitational forces and begin spreading out? All the evidence in our universe demonstrates that, as objects acquire more mass, they become heavier and denser, giving them a higher gravitational field, which holds them even more tightly together. The laws of gravity that affect all matter can't be ignored to explain the initiation of this Big Bang. And secondly, why would this dense sphere of matter exist for a period of time before it suddenly exploded? Did it have a long fuse attached to it?

If by some miracle there was such an explosion, why didn't it immediately collapse in on itself due to the immense gravitational forces contained in all the matter? Scientists move on from this argument about the initiation of the Big Bang and will say, BUT for the fact that we can't figure out how this began, we can postulate how the Big Bang explosion continued to eventually form and evolve into what we see today. That's a big but! This scientific explanation surrounding the beginning makes no sense.

However, I would imagine that if God created the universe, He could have started with an event similar to the Big Bang, or He could have merely placed matter throughout the cosmos as He created it. He didn't need to start it from some central point by an explosion. We shouldn't limit God's ability to create in such a way that makes it easier for us to comprehend how it all got started. He may have decided to create on a grander scale, beyond our imagination or comprehension.

2. Runaway Universe

When the Big Bang theory was first studied, scientists believed that the evidence of this event would show that this outward expansion would slow and eventually collapse into a reoccurring cycle of Big Bangs. In recent years, scientists have been shocked at a new discovery that has left them with a difficult dilemma.

They have found that the further they gaze into the universe, the faster distant objects appear to be traveling away from us. They've calculated the acceleration of distant galaxies within our universe and have concluded that the universe appears to be drifting off into the vastness of space, never to return. This also means that the universe had a beginning and will never collapse into a never-ending cycle of Big Bang explosions. Of course, this finding again is driven by their conclusion that light photons are not affected by gravity. So, these scientists are scrambling to come up with different explanations for these observations other than to go back and challenge the assumptions behind the Big Bang.

Some scientists are now postulating that maybe matter will eventually slow down and return, while others are theorizing that this far-flung matter would enter other parallel universes and add to their overall mass. So I guess we would be having a universal mass exchange program, even though they claim there is a boundary at the edge of our universe, with nothing beyond its border. Forcing unsubstantiated conjectures into your arguments to support a flawed theory actually compounds your initial errors.

3. Distribution from Center

Have you ever studied the science of explosions? Where does all the debris of the explosion go? It's all flung out, leaving nothing behind at the origin of the explosion. Only the gravitational forces on Earth cause a small fraction of an explosion on our surface to fall back down to its point of origin. If you performed this same stunt in space, it would scatter from the explosion, never to return. So why do we find the visible matter in the universe dispersed everywhere with no huge void in the center? If the Big Bang started with an explosion, we would find all the matter at the edge of the universe like the skin on an inflated basketball.

Just look at the Crab Nebula supernova remnant in the Taurus constellation. (See Illustration #9.) This fairly recent supernova explosion occurred only 180 years ago, yet we can see that most of its matter is distributed around its exterior. The only matter remaining within its explosion circumference are solar systems that were initially located outside the initial explosion zone. This explosion pales in comparison to the supposed Big Bang, where all atomic matter would have had an explosive effect, thus accelerating all matter far from its center.

What would be the odds of an explosion in space leaving matter throughout the universe without a massive central void, and at the same time, finding matter expanding at higher velocities the farther it is away from Earth? As we will discuss these odds in our summary of this chapter, they are astronomically improbable.

4. Separation of Matter

If matter was flung out from a Big Bang, would it separate and not form into celestial bodies? This might be especially true at the outer edge of the visible universe where it appears to be rushing away from us and, therefore, must be separating apart as well. Or is it? We could argue that if this matter separates too quickly, it will not form into galaxies. But the further we look into the universe, the more galaxies we see. We are seeing these numerous galaxies at the edge of our universe. We are, however, witnessing them as they appeared 13.8 billion years ago, and I'm sure many of these young galaxies that we observe to be close together would have combined to form larger galaxies by now.

Also note that if the initial explosion of the Big Bang occurred with total separation of all matter, why would it eventually combine again when it was farther apart? Oh, I guess you could theorize that matter slowed down after the initial Big Bang explosion due to gravitational attraction, but that gets us back to the old argument that you would first need to suspend gravity and then somehow cause all matter to repel itself in a massive explosion.

If we were able to view a three-dimensional map of the entire universe as it stands today, taking into consideration that we would need to accelerate in time our view of distant objects, what would we see? We would find a fairly consistent pattern of galaxies throughout the universe, separated by vast areas where voids exist. It's highly unlikely that this would be representative of the remnant left over from a Big Bang explosion.

5. Alignment of Galaxies

As discussed in the previous chapter, we noted the significance of visual displays that illustrate the pattern of vast areas in the universe that seemingly have no matter contained within them. We could conclude that these voids were created by gravitational forces that pulled matter into the surrounding bands of galaxies. These depictions, showing the universe with bands of galaxies surrounding massive voids in space, have sometimes been described as viewing a cosmic web. You'll note in these simulations that galaxies are not evenly spread throughout the universe, but instead are organized into twisting bands containing massive numbers of galaxies. Another way of illustrating this would be to think of the universe as containing clusters of large bubbles, and that these bubbles contain in their interior vast areas of empty space. Galaxies are found in strands or bands, which are attached to the surfaces of these immense galactic bubbles or voids.

The discovery of these large voids separating galaxies suddenly became an obstacle to the timeline constraints of the Big Bang explosion. There just wasn't enough time to form this structure of the universe within the window of only 13.8 billion years, assuming gravity was the predominant force moving matter around in the universe. So cosmologists had to come up with a new idea, and it was a whopper. They concluded that these voids were filled with the invisible dark matter and driven apart with dark energy [30]. Due to the absence of matter in these voids and needing to fit their model of a homogeneous expanding universe, cosmologists needed to fill the voids with something. This theory was concocted with no evidence to support their conclusion. They theorized that these invisible and non-detectable forces were supposedly expanding the voids in space and driving the galaxies apart [31]. They had to reach this conclusion because they knew that only relying on gravitational forces to explain the Big Bang creation could not account for these dark regions of space amidst a runaway universe, while making it all fit within the timeframe of a Big Bang creation window.

Another piece of evidence that launched scientists into this quest to create dark matter was an observation about the rotation of galaxies. You would think that stars located closer to the centers of a galaxy would orbit their individual galaxies much faster than those in distant orbits. But it was discovered that all stars within a galaxy circumnavigate or orbit their galaxies at the same rate. This rotation of stars within galaxies could only be explained if their center core contained massive and invisible matter. They felt that dark matter could explain this observation. Years after their theory on dark matter was proposed, they found massive black holes at the centers of galaxies, which solved their initial concern about the rotation of galaxies. So why haven't they retracted the need for dark matter?

---

[30] "Dark Matter", Science Clarified, http://www.scienceclarified.com/Co-Di/Dark-Matter.html

[31] Nola Taylor Redd, "What is Dark Energy?", https://www.space.com/20929-dark-energy.html

They now say black holes could also be considered dark matter. However, black holes contained within galaxies would not explain why there are large voids in space, so they turned to the mysterious dark energy, which they now believe is driving everything apart at an accelerated pace. As an interesting note, some in the scientific community are starting to doubt the existence of dark matter and dark energy.

So how do you explain the formation of these voids in space where galaxies are arranged in bands or strands surrounding these voids, and why did this become such a roadblock to the Big Bang theory? Well, if the creation of the universe were much older and not expanding, and if matter was initially spread out over the cosmos, then over a longer period of time gravitational forces alone could explain what we visualize in the universe. But this would mean that the universe is much older than the Big Bang creation theory.

Under the scenario which theorizes that the beginning of everything in the universe was already dispersed without the need for a Big Bang, how would we see this type of beginning evolve into what we see today, if we assume that gravity was the only force in the universe? Well, if we went back in time under this scenario, we would find matter had initially been placed throughout the cosmos. It would begin to form larger and larger objects, starting with dust clouds that would form asteroids, which would form stars and planets, which would eventually combine to form galaxies. This process alone would take a very long time. But once galaxies were formed, what happened to them? Galaxies would, of course, attract other galaxies and would be attracted into twisted lines of galaxies where they would slowly attract their neighbors. This process would leave vast regions of space void of matter which had been attracted to the bands of galaxies.

You may ask, why are galaxies arranged in strands or bands? Well, it's fairly simple. As galaxies would begin a process of pulling together in a three-dimensional space, their acceleration toward other galaxies would be influenced by how many galaxies were pulling on them through gravitational forces. If for example, seven galaxies of equal size in a cluster pulled on one distant galaxy, the distant galaxy would be accelerated much faster towards the cluster of galaxies, leaving an area of empty space. But if there were two clusters of seven galaxies and one galaxy equally spaced apart and arranged at opposite points in a triangle, the lone galaxy would be pulled between the two galaxy clusters to form a straight line. This would occur because the single galaxy would have less attractive force and would be accelerated much faster than the two clusters. As the smaller galaxy would be pulled toward the two clusters, each cluster would pull it with an identical force, but instead of being pulled to one of the clusters, the lone galaxy would find itself pulled to a resting point between the two clusters.

Now, let's take that one step further. As galaxies begin to line up into bands along a line, the galaxies within this line would be pulled in opposite directions by neighboring galaxies within the line. The resultant sum of forces on each of the galaxies would be equal, but from opposite directions within the line of galaxies, which would tend to significantly slow these galaxies from being drawn together. But as other galaxies approach this line of galaxies tangentially, they would be accelerated toward the line of galaxies, which would pull them within the line, where they might reach a somewhat stable footing.

Now shift your thinking from two dimensions to that of three-dimensional space. These lines of galaxies would not be located on a straight line but would instead find themselves in a band of galaxies that twist and turn and wrap around large voids in space. We can see the early stages of this clustering of galaxies into bands when we look at deep space. There we see thousands of smaller and younger galaxies spread out before they have begun to form into larger galaxies within bands. I wish we could speed up their motion and see if they have begun the process of combining to form larger galaxies and aligning themselves into bands. Just think about how long all of this would have taken.

Also, just think how long it took to develop the massive black holes at the center of galaxies. The Milky Way galaxy, which they believe has nine massive black holes, must have been formed from the collision of tens of thousands of smaller galaxies. Mind boggling! But think how long this would have taken. I would note that this is also evidence of a beginning and an end to our universe, however big it is. Scientists now agree that the universe will come to an end when everything coalesces into black holes and goes dark. So much for the theory of a runaway universe that's expanding.

There is no way that the formation of voids in our universe and the creation of large galaxies fit within the timeframe for the Big Bang. The forces of gravity, and a much older universe that is not expanding, prove to be the simplest and most logical explanation for the distribution of matter in space.

6. Evidence of Coalescence

The Big Bang theory presupposes that light photons are slowing down from an expansion of the universe. But when we look throughout the universe, we see the opposite occurring. We see that everything in the universe is, in fact, coalescing together, even at the edge of the known universe. For example, if the Big Bang had occurred, we would not be seeing galaxies at the edge of our universe, as all matter would be accelerating apart and too diffuse to form objects.

As discussed earlier, gravity is the primary force that has pulled everything into galaxies, having black holes as their central combiners. This leaves large areas of space void of any matter.

I would also argue that all matter contained within galaxies, even though it may be flung out from the center, never leaves the gravitational pull of that galaxy. We know the end point for our universe, if it were to continue to coalesce, would be an existence of total cold darkness containing massive black holes. So if the universe is presently coalescing, how can it be expanding? The two can't exist simultaneously.

7. Age of the Universe

The age of the universe under the current thinking surrounding the Big Bang theory is a moving target. Scientists began their assessment of the age of the universe by first determining how far away they have observed objects at the edge of the universe. This can be accomplished by calculating the distance to other galaxies through observing the brilliancy of Class 1-A supernova explosions contained within these galaxies. They then calculated how long it would take for light photons to reach us here on Earth from these distant galaxies. The result was 13.8 billion years. But wait, how long did it take for the galaxies we see at the edge of the universe to reach their position? The 13.8 billion years is negating the time needed to expand matter to its current observed distant location in space. So, assuming the Big Bang hurled debris from the central explosion at an average velocity comparable to the speed of light, it would have taken another 13.8 billion years for matter to reach where we see it today. So maybe the universe is now 27.6 billion years old [32].

---

[32] Ethan Siegel, "If The Universe Is 13.8 Billion Years Old, How Can We See 46 Billion Light Years Away?", https://medium.com/starts-with-a-bang/if-the-universe-is-13-8-billion-years-old-how-can-we-see-46-billion-light-years-away-db45212a1cd3

Like a Ginsu commercial for kitchen knives would proclaim, "But wait, there's more." As discussed earlier, the Hubble telescope has recently observed galaxies some 30 billion light years away. But wait, there's more. They've thrown an effect caused by dark matter and energy that would make you believe the edge of our universe is now 41.4 billion light years away [33]. But wait, there's more. They believe that they have found background radiation left over from the Big Bang some thirty billion light years away. These last two findings are interesting when trying to find the edge of the universe, but this same evidence could also provide proof of a much wider and maybe infinite universe that may have been created instantaneously.

So realistically, how old is our universe? Under the current Big Bang theory if we assume the universe is 27.6 billion years old and it takes 250 million years for the Milky Way galaxy to make one full rotation, our galaxy would have only rotated 110.4 times in its entire existence to fit the Big Bang model. It must have taken longer to form our galaxy. The closest galaxy to merge with our Milky Way galaxy is the Andromeda galaxy, which won't occur for another 5.2 billion years. It will probably take several more billions of years before these combined galaxies become concentric again.

Our galaxy has gone through this process of merging with other galaxies many times before. Recent discoveries

---

[33] Ethan Siegel, "If The Universe Is 13.8 Billion Years Old, How Can We See 46 Billion Light Years Away?", https://medium.com/starts-with-a-bang/if-the-universe-is-13-8-billion-years-old-how-can-we-see-46-billion-light-years-away-db45212a1cd3

show the Milky Way galaxy has nine massive black holes. This is evidence that at least eight other massive galaxies having combined with it in the past, and who knows how many countless globular clusters and small galaxies have combined to make up its massive breadth. And how long did it take for matter to form into nebulae and stars, and for stars to burn out over billions of years to begin the formation of black holes? Also consider that our galaxy is not the biggest one out there. We could very well envision the universe's age being in the trillions of years old, not merely 27.6 billion years old.

However, we do know there's a beginning to all of this, and we know there's an end point as well. But we also now know by our observations that it's extremely old.

8. Gravitational Effects on Light

The one and only piece of evidence that led Hubble to believe the universe was expanding was his discovery that light photons appeared to be slowing down the further they traveled across the universe. But instead of accepting this as fact, Hubble chose a different conclusion based on his assumption that light photons couldn't possibly be affected by gravitational forces, and therefore he theorized the universe was expanding. However, the recent discovery that gravitational waves in space travel at the speed of light provides proof that gravitational attraction has a limit, as we discussed in Chapter 6. This new theory formulates that the speed of gravity has a dampening effect on approaching objects that travel at speeds approaching gravity's pulling velocity. Since this speed of gravity is equal to the speed of light, we would notice a reduction in the velocity of light photons traveling across the universe.

This new theory about gravity slowing down light photons would help resolve several dilemmas facing scientists about their Big Bang theory. Namely, the age of the universe relative to the apparent time needed to coalesce the universe into what we see today. There would be no need to invent dark matter and dark energy to fudge the Big Bang timeline. This would also resolve the concerns of a runaway universe that would never collapse. This would also prove once and for all that there is a link between science and creation, albeit a very old creation. It would also get past the elephant in the room concerning how the Big Bang could have begun.

9. Gravity Waves

Recent discoveries have shown that gravity waves exist and travel at the speed of light. Additional research is needed, but it appears that these waves are not affected by other forces. If it is determined that these waves travel from the edge of the universe and arrive here at the speed of light, this would prove that the universe is not expanding but only coalescing.

10. Absence of Light Following Celestial Events

The recent discovery of gravity waves has shown that they travel at the speed of light. What is interesting about these events is the fact that cosmologist's have not detected or witnessed visible light being emitted from these gigantic collisions. This is evidence that visible light has been delayed in transit, and most likely delayed due to gravitational forces.

11. Size Matters

One of the tantalizing offshoots of the Big Bang theory was that scientists felt they could better explain why all matter was identical in size and function for all atomic particles by merely noting it all came from the one and only singularity and therefore was identical. Sounds nice, even though they can't get past how the Big Bang started. This also allowed science to distance itself from a creation theory that made it harder to explain why all matter in the cosmos is identically sized, with identical functions, when it was instantaneously placed throughout the cosmos.

12. What Else Could Affect Photons?

Scientists have also been toying with other ideas that might explain why light photons slow down in space.

- **Higgs Field**

    We briefly discussed the Higgs Field effect earlier in this book.

    Higgs was trying to derive a theory that would explain why they have observed some particles in space slowing down. Could this be due to gravitational forces, or maybe the bombardment of subatomic particles? There isn't enough research completed in this area to draw any conclusions at this time.

- **Quantum Physics**

    The theory of quantum physics or quantum mechanics attempts to calculate mathematically the mechanics and forces behind atomic and subatomic matter. This area of research is totally theoretical, but if there is any validity to it, there could be found a relationship between photons and subatomic particles which could account for a deceleration of photons over time. I have some reservations with this possibility from a perspective that if you were to bombard individual photons that are traveling together from a distant star with any kind of particle, you would see a scattering effect of light as these photons traverse the universe. This would result in making distant objects appear out of focus.

    On the other hand, if you apply gravitational forces evenly across an array of photons traveling in space, they could be slowed while staying together and not scattering.

An example of gravitational effects on photons which changes their direction but allows you to see clearly that the photons stay together is found when observing an Einstein Ring. (See Illustration #4.)

13. Energy Depletion of Light Photons

When light photons are created and leap off their illumination source, like a flashlight filament, the photons instantaneously begin traveling at the speed of light. This would be like a bullet leaving the muzzle of a rifle. But then its electromagnetic wave supposedly becomes the energizer bunny, never to run out of energy and never to slow down its companion photon particle. Why is that? I understand they are traveling in the vacuum of space, where you might think there is nothing else to influence them. Of course, this author believes gravity plays the primary role in slowing down photons. But what if photons themselves lose some of their kinetic energy, or their electromagnetic wavelength elongates giving it a redshift appearance of slowing down when it's traveled for billions of years.

Scientists have known for years that particles like microwaves loose some of their kinetic energy when traversing the universe. This energy reduction of electromagnetic particles has been adopted within the scientific community. There are those in the scientific community that have calculated the depletion of kinetic energy relative to light photons and believe that gravitational forces in the universe can reduce the kinetic energy of these particles over an extended period of time. They believe that this reduction in energy may not decelerate the photons, but that it would result in the elongation of their electromagnetic wavelength. This elongation would give the appearance that light photons are redshifted or moving at lower velocities. This finding could be used to counter Hubble's theory of an expanding universe.

As explained in Chapter 6, we could also find that gravity could be performing a dual role in the deceleration of light photons. Gravity could be reducing the velocity of light photon particles as well as elongating their electromagnetic wavelength. Remember that missing from the massive cosmic collisions were any evidence of a bright flash of light photons, which could have been decelerated by gravitational forces over time.

14. Center of the Universe

One of President Ronald Reagan's most famous and often-used debate lines was, "Well, there you go again." We have seen throughout history that mankind has tended to place itself as a central focal point in its perceived cosmos. Call it a search for significance. We see this same perspective being played out in the Big Bang theory where they proclaim that the universe is expanding away from a central point. But where is this central point? Interestingly enough, astronomers peering into deep space in all directions see what appears to be the edge of their universe some 13.8 billion light years away. So where would you suppose Earth is located? Well, it's in the center of this Big Bang! "Well, there you go again." Scientists don't want to admit it, but it's very obvious. So why are we again at the center of our universe? Mankind has always dreamed of being at the center of his universe, so why should we be so surprised?

Not surprisingly, scientists recognized this concern about being centrally focused, so they came up with a deflecting argument. Some have tried to argue that the expansion of the universe is like the expansion you see when baking a muffin. This theory notes that if you were a molecule anywhere within a muffin, you would see the same expansion taking place around you when the muffin expands while it's being baked. But if you believe the muffin has a limited size, and you are able to see the edge of the muffin and then note that you appeared to be in the center, then you're in the center. Kind of half-baked!

So are we truly at the center of our universe or just not seeing far enough into the cosmos to understand our insignificance relative to an infinite cosmos?

15. Refocus on Infinity

Supporting the Big Bang theory is becoming sketchier, which has caused some cosmologists to refocus on an infinite universe again. I think this is primarily driven by the realization that if they only assumed there was one Big Bang universe, then that would be insignificantly small in the infinite cosmos. A second reason might be that as they extend their search into the heavens, they find more objects of interest out there and are coming to the realization that this universe may not be finite. So we're seeing support for a single finite universe now being replaced with a multiverse scenario. But this multiverse scenario, which envisions multiple Big Bangs, carries with it the same flaws we've discussed. A more logical conclusion would be that the infinite cosmos is merely filled with infinite matter, not separate pockets of matter.

Conclusion

I commend Lemaitre's desire to find a link between science and religion. I also know that people have grasped at the Big Bang hoping it would provide them with evidence of God's creation, but this theory has more holes in it than a 200-pound wheel of swiss cheese. Besides, there might be a more realistic nexus between science and religion that we have yet to discuss.

I'm also noting that other cosmologists are beginning to doubt the validity of the Big Bang theory. It's safe to say that if gravity were found to slow down light photons and elongate their electromagnetic wavelengths, and/or if they find that gravitational waves don't slow down as they traverse the universe, this would be concrete proof that there is no expanding universe, Big Bang, or multiverse—just an infinite cosmos with infinite matter. So is there another possibility to explain the existence of the cosmos?

We'll continue this debate in the next chapter by analyzing the evidence that's been gathered so far to determine the most probable creation theory.

ILLUSTRATIONS

Illustration # 1: Sleeping Under the Stars

Remembering camping trips and sleeping under the stars in the Sierra Nevada mountains during my youth.

Illustration # 2: Galaxies in Deep Space

Deep space image of early galaxies at a distance of thirty billion light years.

Illustration # 3: Andromeda Galaxy

The Andromeda Galaxy is one of the closest galaxies to the Milky Way. There are no noticeable spiral arms extending out from this galaxy.

Illustration # 4: Einstein Ring

The distant galaxy is distorted by the gravitational forces from a single sun as the galaxy's light photons pass near to the sun. This distortion makes the galaxy appear to be stretched out in the shape of a ring. This provides proof that gravity can affect the motion of light photons traveling in space.

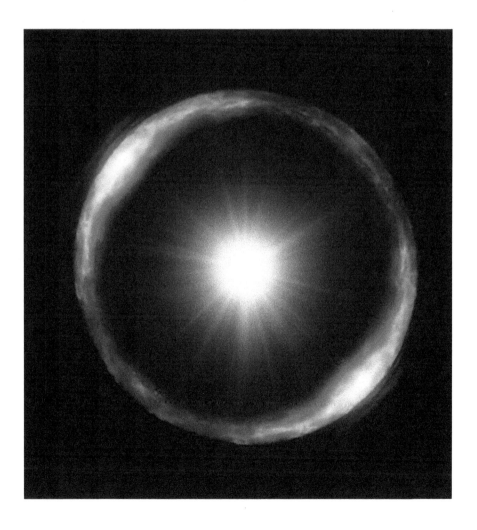

Illustration # 5: Electromagnetic Spectrum

This diagram illustrates the full spectrum of electromagnetic particles. Visible light makes up a very small portion of this spectrum. The redshift effect noted by Hubble detected the longer wavelength of visible light particles traveling across the universe.

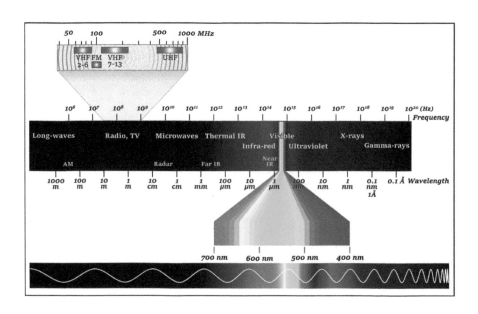

Illustration # 6: Black Hole

Illustration of starry disk around supermassive black hole. Notice that the black hole is hidden in a dense cluster of stars at the center of the galaxy.

Illustration # 7: Nuclear Binding Energy

This graph depicts the binding energy per nucleon vs the number of nucleons. This illustrates that the binding energy for hydrogen is very low. When hydrogen is fused the result will yield a higher binding energy per nucleon which will release excess energy like what we see on our sun's surface.

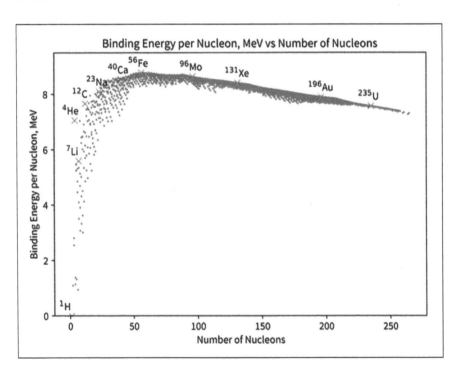

Illustration # 8: Life Cycle of a Star

In this diagram suns begin their existence in the Stellar Nebulas. At the end of their life, they become smaller and denser. If they aren't transformed into black holes, they will eventually merge with other dense objects and become black holes.

Illustration # 9: Supernova Explosion

The Crab Nebula supernova remnant in the constellation Taurus. The debris from the explosion is concentrated around its outer perimeter.

Illustration # 10: White Dwarf

Planetary nebula with a white dwarf in the center. The nebula could have been formed from the ejection of material during the formation of the white dwarf.

Illustration # 11: Black Hole Illumination

A black hole in space being illuminated by a distant cluster of stars.

Illustration # 12: Black Hole – Lens Effect

Black hole with gravitational lens effect in front of a bright star. Provides evidence that light is redirected by black holes.

Illustration # 13: Neutron Star

A highly magnetic rotating neutron star.

Illustration # 14: Great Globular Cluster

A dense cluster of millions of suns that are being held together by their combined gravitational forces and orbit around its central point. Located near the Milky Way galaxy this tightly packed cluster of stars is one of the largest detected clusters. It most likely has a black hole at its center.

Illustration # 15: Spiral Galaxy

The bent arms of this spiral galaxy have ejected solar systems away from its center. Over extended periods of time this material will be drawn back into the gravitational pull of the galaxy.

Illustration # 16: Concentrated Galaxy

There does not appear to be any spiral bands emanating from this galaxy. This could be influenced by a higher concentration of massive black holes at the center of this galaxy. This galaxy might have also avoided mergers with other galaxies long enough for its spiral bands to have been reacquired into a concentric circle.

Illustration # 17: Galaxy Merger

The merger of two spiral galaxies. Notice the far-flung spiral bands that will eventually be reacquired.

Illustration # 18: Voids in the Universe

Glowing lines of detected flare plasma in the universe which are emanating from distant galaxies provide evidence of massive voids in space with galaxies spread out in twisted bands. This is similar to the images derived in the Millennium Simulations reference in the Bibliography.

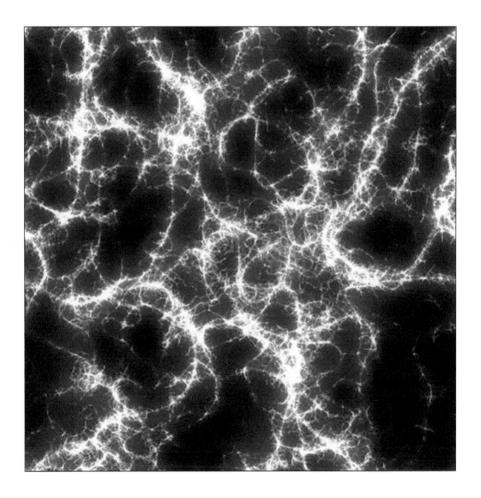

Chapter Ten

Just the Facts, Ma'am!

Finding the truth is as simple as following the facts.

DO YOU REMEMBER the old TV series entitled *Dragnet*? This show featured two police detectives going about their daily routines, attempting to solve crimes on the streets of Los Angeles. Inevitably in each episode sergeant Joe Friday would ask a female witness to give him, "Just the facts, ma'am, just the facts." These detectives knew they had a better chance of solving their cases if they focused strictly on factual information and didn't allow their witnesses to provide hearsay information, or ramble on about matters that had no bearing on their case.

In this chapter we will focus on factual information—just like the detectives—to attempt to solve the mysteries of the universe. Thus far in our search for the truth about our cosmos, we have found several pieces of factual information we can utilize to sort out which scenario best fits the reality of our cosmos. So what do we now know about the universe?

1. **All Matter is Made Up of the Same Atomic Particles**

   All the visual observations of matter in our universe support the finding that everything is made up of identically sized protons, neutrons, and electrons. There is no dark matter, anti-matter, or worm holes in space.

2. **The Perfect Design of Particles**

   The perfect design of the atomic structure allows for the formation of different elements. These elements can't be changed easily. The fusing or splitting of atoms yields energy releases that fuel everything from the warming of planetary cores to igniting the fusion reactions necessary to sustain the flow of energy from the billions and billions of suns.

3. **Gravitational Forces at Play**

   All evidence supports the fact that gravity is the primary force that drives the motion of all matter within the cosmos. Gravity's incredible design allows this force to attract other objects, which combine to form everything from asteroids to black holes that form the centers of massive rotating galaxies.

4. **Gravitational Wave Velocity**

   Current evidence has detected the velocity of arriving gravity waves at the speed of light even though these waves have been generated far away. The distant origin of these events should have caused a deceleration of these gravity waves had they occurred in an expanding universe.

5. **Gravitational Effects on Photons**

   Gravitational forces can affect the position of light photons traveling through space as evidenced in the Einstein Ring phenomenon. These same gravitational forces can also elongate the electromagnetic wavelength of light photons, giving them the appearance of being redshifted, or traveling at a reduced velocity. We can therefore assume that gravitational forces will continue to interact with light photons as they travel across the universe.

6. **Visible Light Delayed**

   There is an absence of visible light arriving from distant cosmic collisions. This provides evidence that light photons decelerate due to gravitational effects as they traverse the universe.

7. **The Universe Is Coalescing**

   It is very apparent that wherever we look in the universe, we see matter coalescing into larger celestial bodies, which eventually combine to form galaxies. The scientific community agrees that galaxies will be combined and eventually be absorbed by their massive black hole centers, which will leave the universe, cold, dark, and void of any visible matter. There is no evidence that black holes, which are so dense, would ever reach a point where they would defy the laws of physics and explode. One could call this the end of the universe, but not the end of space.

8. **The Universe Has a Beginning**

   If the universe has an ending, then it must have had a beginning. If we went back in time and unwound the motion of everything in the universe, we would unwind the formation of everything. We would find ourselves back at the beginning of the universe. This is evidence that the universe was created in some form or fashion, and therefore was not in existence for all eternity.

9. **Expanding Boundaries**

   The boundaries of our universe keep expanding as scientists are able to peer deeper into the cosmos. So why wouldn't we find matter stretching further into the cosmos as we increase our ability to look further into space? This doesn't mean the universe is expanding. It simply means we would be able to observe matter in space further than we've seen before.

10. No Reoccurring Big Bang

You can't claim to have an expanding universe that will eventually collapse and repeat itself when the evidence you have relied on now supports that there is a runaway expansion of the universe. Correct the flawed assumption used to theorize an expanding universe.

11. The Universe Is Older Than the Big Bang

The distribution of matter in the universe points to a much older universe than what would fit within the Big Bang theory. The existence of large voids and the distribution of galaxies within the universe illustrate that the forces of gravity were at work over a much longer period of time than that needed to support the Big Bang theory.

12. Lack of Evidence for Dark Matter Existence

Inventing dark matter and dark energy in an effort to fill in and expand the massive voids in the universe, with no evidence of their existence, is grasping at straws instead of trying to search for the real truth. When black holes were discovered to be at the center of galaxies, this evidence alone negated any need to explain the movement and rotation of galaxies by inventing some other force or matter.

Analyzing the Evidence

This completes our discussion about the factual evidence which supports the foundation for the creation of the universe. From this brief synopsis, we can conclude that the universe has a beginning and an ending, and that it's made up of identical matter that is uniquely designed. It is also much older than the Big Bang and extends beyond our ability to detect.

There still remains one area that needs to be fully resolved. We need to determine whether the universe evolved from a single creation consisting of diffuse matter, as discussed in Genesis, or whether the universe was created in a Big Bang explosion. I believe that I've provided overwhelming factual evidence that eliminates the creation of the universe from the expansion of a singularity. There are several in the scientific community that agree that the Big Bang theory is highly improbable.

None the less, let's go through the exercise of tabulating assumptions for each of these theories. In Chapter 3 of this book, we set up evaluation criteria to help solve this puzzle. There we noted that we would first rely on physical evidence to make determinations. But where we found insufficient evidence to make a solid recommendation, we would utilize Occam's Razor as a tiebreaker. Even though this exercise is somewhat irrelevant as there is sufficient evidence that eliminates the creation of the universe from a singularity, let's determine which theory is supported by the least complicated and least number of assumptions?

Just the Facts, Ma'am!

Even though I feel that I've provided ample logic to support the theory that gravitational forces slow down light photons, I've chosen to list this as an assumption for sake of our analysis. Along with this, I've assumed there is currently insufficient factual evidence to verify if gravity waves can disprove the expansion of the universe. I've also listed, as an assumption, Hubble's theory that light photon deceleration is evidence of an expanding universe.

To support the Big Bang theory, you must assume the following:

1. That the initiation of the Big Bang started with the elimination of the laws of physics relative to gravitational forces, which no one has been able to explain.

2. That the outward acceleration of matter following the Big Bang explosion would not leave a large void in its center or disperse most of the matter at the far reaches of the Big Bang explosion.

3. That dark matter and dark energy exist and fill the large voids in the universe even though there is no evidence of their existence.

4. That light photons are not decelerated by gravity but are instead evidence of an expanding universe.

5. That distant gravity waves will be detected on Earth at reduced velocity below the speed of light, as evidence of an expanding universe.

6. That the universe is only twenty-seven billion years old despite the extensive time needed to form all the galaxies and separate them throughout the universe.

7. That the expanding runaway universe would eventually reverse itself, resulting in reoccurring Big Bang explosions.

8. That the Big Bang universe is finite in an infinite cosmos.

To support the Genesis Creation theory, you must assume that:

1. Light photons traversing the universe slow down and have their electromagnetic wavelengths elongated due to gravitational forces. This gives them the appearance that the universe is expanding, when it's actually at a steady state, neither expanding nor contracting, but only coalescing.

Conclusions

When assessing the evidence and using Occam's Razor to sift through the assumptions we find overwhelming support for the Genesis Creation as the most logical option behind the creation of the universe with the least number of assumptions to support this conclusion.

The probability of a Big Bang expansion is therefore extremely remote, even when you negate the theory about gravity impacting the travel of light photons traversing the universe. This new postulation that gravitational forces slow down light photons and elongate their electromagnetic wavelengths, brings the realization that the universe is not expanding, but is only coalescing, and is influenced solely by the forces of gravity. This theory may be reinforced when additional studies are conducted on gravity wave velocities.

If Hubble had focused more on the true meaning behind his observations of redshift, which became the sole basis for his theory that led him to conclude the universe was expanding, we might have come to an earlier consensus that we are, in fact, a part of the Genesis Creation.

The only remaining question is, how far does matter extend throughout the cosmos? We could find that matter in the universe extends forever, or we could find that God set a limit on how far He chose to extend His creation. I believe, but can only theorize, that the cosmos must be infinite. However, God only knows the answers to these questions, and man shouldn't limit His creativity.

Chapter Eleven

What's Beyond Our View?

Twinkle, twinkle, little star, what's beyond you way out far?

THEORIES ABOUT WHAT we might find out there in the distant cosmos are a dime a dozen. They range as far as your imagination, from the existence of parallel realities, to string theories, to multiverses, to wraparound universes made up of all shapes and sizes. These tantalizing theories are full of strange and wonderful possibilities. But we should also consider the possibility that the cosmos is filled with an infinite display of matter, just like what we currently see in our universe.

Our review of history has shown us that we tend to limit our perspective of the universe to what we can see and touch. We have seen that as mankind expanded his ability to peer into the cosmos, he has found the edge of his universe expanding with it.

I once worked for a power plant manager who had a philosophy about work ethics. He noted that workers tended to pace themselves to complete work assignments within the scheduled timeframe. In attempting to improve productivity, he would use this statement: "Work expands to the time allowed." Sure enough, whether you set a relaxed or more realistic schedule for a group of employees to complete a time-consuming project, the project managed to be complete at the end of the schedule. So this manager focused on setting aggressive but attainable work schedules and held people to those timelines for the completion of their work. I believe we can apply this philosophy in a parallel fashion to explain the current perspective of the cosmos. We could state it as, "The breadth of our universe, expands with our ability to envision it." So are we limiting our perspective of the cosmos to what we can see, as humanity has done in the past? As we expand our view into the cosmos beyond its current reach, it will be interesting to find out whether we see the continuation of this universe or something else.

We should note that if the universe or cosmos had a beginning, which all evidence points to, then light photons from the outermost reaches of space may not have reached us yet. This might give us the illusion that matter is nonexistent in space beyond a certain point. We should keep in mind that matter might still be there. I'm looking forward to newer telescopes that will peer even further into space, allowing new discoveries beyond what our current devices can detect.

Also, when we are projecting what might exist beyond our ability to detect, shouldn't this distant existence match the same patterns for what we see in our universe? It would make no sense that a different reality would form in different regions of the cosmos. With us proving that the expansion of our universe is extremely unlikely, the most logical model for our universe is the continuation of matter as we currently observe. It would make logical sense that this matter extends forever.

We should also note that several of the theories about matter in the cosmos, like the multiverse theory, which assumes multiple existences of Big Bang universes, are stuck in an illusion that our current universe is limited in size. If matter were spread throughout the cosmos, why would it be restrained to multiple runaway universes, and not be spread out as we see in our universe? If the effects of gravity on light photons provide evidence that our universe is not expanding, then the most logical conclusion would be that we are part of an infinite cosmos with infinite matter, from a Divine creation. If not, that would be a tremendous waste of space!

Chapter Twelve

Six Days or 6 Days?

To understand His Creation, you must focus on The Everlasting.

ONE OF THE long-standing debates about God's Creation within the religious community is whether it occurred over a span of six calendar days or, if it occurred over an extended period of time. This debate is sometimes referred to as a debate between New Earth vs. Old Earth. To help resolve this debate, let's again look at the Biblical description about how our universe began. Genesis 1:1-2 states, *"In the beginning God created the heavens and the earth. And the earth was formless and void, and darkness was over the surface of the deep: and the Spirit of God was moving over the surface of the waters."* In the first part of Moses' account of creation, his description seems to denote that all matter was created in the cosmos and spread out in such a fashion that there was no form to it, and therefore there were also no planets, no suns for light, and no distant galaxies. From this beginning, how long did God move over His creation and set in motion the formation of the stars and planets?

I would first and foremost point out that no matter which theory is correct, the bottom line is that there is no other explanation for the existence of the cosmos other than to say that God created the cosmos and everything in it. After our review thus far, we can state this as factual, knowing that all the evidence points to a cosmos having a diffused beginning and everything in it having an identical design.

I would also like to make another point to the religious community. We should be careful not to adopt theories about the creation of the universe that are filtered through man's limited ability to perceive and predict what may have started this creation. Therefore, this discussion will focus more on two possible theories and their likelihood of occurrence, but by no means can we fathom all the mechanics, processes, or timeline possibilities that God may have chosen to form and develop His creation. With all that being said, let's explore these two creation theories and see if we can select the more probable theory.

The Creation Occurring in Six Calendar Days

This New Earth creation theory focuses on the literal translation of Moses' account in the book of Genesis. This theory assumes that this creation occurred around 6,000 years ago, based primarily on the genealogy records contained within the first few books of the Bible. The initial creation in this account notes that, *"God created the heavens and the earth,"* as stated above in Genesis 1:1-2, but that, *"the earth was formless and void, and darkness was over the surface of the waters; and the Spirit of God was moving over the surface of the waters."* Here it sounds like Moses is simply stating that, first, everything was created in this initial creation and there was no form to matter.

It's also generally believed in this New Earth option that the first day of creation encompasses the creation of the cosmos. It describes all matter being created but, interestingly enough, does not indicate if matter in the universe had taken form as planets, suns, or galaxies. We only know that at this time the earth was formless and void and without sunlight. But I think we can also conclude that if the earth had no shape in this initial creation, this could also imply that everything else in the creation started its existence in a similar state of formation.

We next see a reference in the second verse of Genesis 1 that, *"The Spirit of God is moving over the surface of the waters."* So as He moved over this existence, God caused the matter to form into planets and suns. The question is, how long did it take for God to first create matter and then form it into what we see in the universe? If God had the ability to create everything in existence, He most certainly had the power to do it all at once. But if God is timeless, why would He want to create what we see today all at once?

This brings up an interesting possibility. With the Spirit of God hovering over this initial creation described in the first two verses of Genesis, is this describing a period of time that may not be linked to the first day of the six days of creation mentioned in the following sections of Genesis? God's Spirit may have taken some time to form and place everything in the cosmos prior to forming our sun and the earth. This could mean that all of the perfectly designed atoms were placed throughout the cosmos in one instantaneous creation, but that God took some time to move about and cause the formation of everything we see today. After that prolonged beginning of the creation, he could have then formed his creation in six calendar days.

In early Biblical times, the universe was perceived as being close at hand. In Moses' account, it's not clear if he perceived Earth as the center of everything. However, it leaves the impression that the heavenly bodies were not very far away. When early man was focused on what they could see with the naked eye, heavenly bodies didn't seem so far away. Moses could not have perceived the vastness of space and all of the celestial bodies at play. We can now understand why the Genesis account focused more on our solar system and its formation. It's also easier to accept and believe that all matter in the heavens, which were viewed to be nearby, could be created and formed on day one.

Evidence that Challenges the "New Earth" Creation

But if God created the entire cosmos, that potentially goes on forever, in twenty-four hours on day one of his creation, how do we explain the following cosmic evidence we see that points to an "Old Earth" creation of the cosmos?

- **Observation of Distant Objects**

    Since we can measure the most distant galaxies at 13.8 billion light years away, we know that it takes light photons that same amount of time to reach Earth. It might be that we just haven't built a telescope large enough to see whether celestial bodies might exist even further into the cosmos. So how could these light photons, traveling for billions and billions of years, only be 6,000 years old? Or another way of looking at it is, how could these light photons have reached us if they have only been traveling for 6,000 years? We wouldn't be able to see most of the stars in our universe or our Milky Way galaxy if the universe were only 6,000 years old.

There is a possibility that if God had the ability to instantaneously create the universe, He could also have made it appear old, so we could see further into His creation. This would mean that He would have needed to set everything in perfect motion and alignment. Galaxies would need to be created with all of their stars set into motion, rotating around their centers. The universe would then need to be filled with light photons and gravitational forces being placed all along the pathway from distant suns and galaxies, traversing trillions of miles, so we could see these distant objects and assume these particles and forces were generated from the source, when they would in fact only be replicated images of distant objects and forces. These streams of photons traveling across the universe would also need to change the image they project, as these distant objects would have naturally changed over time. Again, under this New Earth theory it would make the universe appear old when it really wouldn't be.

- **Gravitational Forces**

    One only needs to gaze at the Milky Way galaxy on a clear moonless night, away from city lights, to witness the immense gravitational forces at play containing all of the stars within its grasp. Now also imagine similar forces across the galaxies that are constantly shaping our cosmos. These forces reach out everywhere across the cosmos and travel at the speed of gravity. When you look at large objects like galaxies that take a quarter of a billion years just to perform one rotation, gravity's influence seems very slow. On this grand scale, it makes everything seem like it's standing still, yet gravity works over billions and billions of years. We can now detect changes in gravitational fields from distant celestial events that take billions of years to reach Earth.

    So again, how could these forces be at play if they have only been in existence for 6,000 years? These forces would have had to be placed throughout the cosmos during the instantaneous creation to give us the illusion that our universe is very old.

Let's continue to investigate evidence that challenges the New Earth theory of creation, but we'll now focus on what we observe here on Earth. It seems like everywhere you explore on Earth you see evidence that points to an Old Earth creation. Here are a few examples:

- **Carbon Dating**

 Carbon dating, which measures the rate of radioactive decay from organic carbon, is predictable and fairly accurate. For example, it places the dinosaur extinction at about sixty-four million years ago, which just happens to be at the same time as the last great meteor impact, which was also discovered through carbon dating. You can debate the accuracy of this science, but there is quite a difference between sixty-four million years and 6,000 years. It's not that inaccurate. I have heard about the claim that they have found what looks like human footprints next to those of dinosaurs. But it's also more probable that this is a clever tourist attraction which actually depicts something entirely different. Though maybe it's evidence that God walked with the dinosaurs.

- **Plate Tectonics**

 Continental plate movement is a very slow process that has shaped and reshaped this planet's surface for billions of years. It lifts mountain ranges and causes some of these plates to slide under one another. There is substantial evidence that these plates have been combining and separating for billions of years.

- **Erosion**

    Extensive erosion has transformed mountains into sediment, which is transported by water into lake beds and ocean bottoms. Over millions of years, these deposits are transformed into sedimentary rock formations, only to repeat the process when new mountains are formed through tectonic plate action. Some have felt that all erosion occurred during the Great Flood depicted in Genesis. But this would not explain the fact that sedimentary deposits are found in rock structures we see today, which take a long time to form and therefore provide evidence that the Earth is extremely old.

- **Fossilization of Organic Material**

    The existence of both ancient plant and animal life has been discovered across our globe. Fossils dating back hundreds of millions of years are found throughout different layers of sedimentary rock. Through carbon dating (or by just noting how long it takes to form this sedimentary rock) we can prove that these remains are evidence of life from a distant past.

- **Polar Shift**

    By examining the magnetic polarization of rock in the Mid-Atlantic Rift zone and noting how long it takes for this rift to spread apart, scientists conclude that our north and south poles flip polarity about every nine million years. They have found evidence of this occurring multiple times.

- **Volcanic Eruptions**

Volcanic action has helped shape this planet. Volcanic action has been at play since the beginning of this planet. As it cools underground, it forms granite. For example, the entire Sierra Nevada mountain range in California is made up of granite that formed underground, then was pushed upward to its current location. When volcanic lava breaches the surface of the earth, it flows downhill, solidifying into solid formations. We see towering mountains and entire islands formed by this slow process over extended periods of time. We also see evidence of erosion on these volcanic formations and note that most of these formations are extremely old.

The Creation Occurring Over an Extended Period

In this Old Earth creation theory, one assumes that Moses was describing periods of time, instead of actual twenty-four-hour days. As discussed earlier, the notation in the first few verses of Genesis where the Spirit of God was moving over the creation, could lead us to conclude that this hovering occurred over some period of time. Under this theory, one would conclude that in the first period, God created the cosmos and all the matter in it, which could have occurred instantaneously or over a period of time. As God's Spirit moved over His creation, He then caused the formation of celestial bodies over a longer period of time. This extended creation would provide enough time for us to see ancient photons of light that have traversed the universe, as well as observe the gravitational interactions between celestial bodies that we see scattered throughout the universe.

But under this scenario, we would conclude that God never stopped miraculously shaping His universe, even to this day. Why would we assume that God would take his hands off His creation once He started it in motion? We pray for God's supernatural interaction in our lives, so why would we believe He ever stopped interjecting into His creation.

The same would apply with the miraculous creation of life, which is so complex that it couldn't have started by random chance, or even evolved into more complex creatures though natural selection. We'll discuss this further in another chapter.

I've also heard this Old Earth theory of creation described as Progressive Creation. This can be stated in simple terms—after God made everything, He never took His foot off the gas pedal of molding and changing His creation. In this analogy, God caused the formation and existence of what we see throughout the universe and all around us on this miraculous planet over an extended period of time.

Progressive Creation does not mean that evolution is responsible for what we see today. Random (natural) selection could never have naturally created and evolved into what we see in God's nature. Even Darwin concluded that if biology were found to be more complex than what he knew at the time, which it is, he fully admitted that the only explanation for today's life-filled planet would be that of a progressive creation. All around us we see the miraculous hand of God, who first designed this existence, caused it to take form, and is continuing to mold His creation.

There are other examples in the Bible where God chose to speak in periods of time. For example, Daniel received a vision from God as noted in the 9th Chapter of Daniel, which predicted when the Messiah, Jesus Christ, would enter Jerusalem to be crucified for our sins. The time frame for this prediction was stated as seventy of seven weeks minus one week. However, in reality, the period of 483 weeks turned out to be exactly 483 years. This may have been done on purpose to disguise the coming of the Messiah, but it is interesting that God chose a different interval of time for the fulfillment of this prophecy. This different description of time intervals, however, does not mean that the Bible is in error.

Another interesting point to make about this Old Earth, or Progressive Creation theory, is that this option is in alignment with our understanding of scientific facts and evidence about the universe and everything in it.

Discussion and Conclusions

If we were to use Occam's Razor to determine the more probable option, we would choose the Old Earth theory of creation. In this theory, all the visual and physical evidence supports the existence of an older universe.

For us to believe in the New Earth option for creation, we would need to assume that God created a universe in a very short period of time, with all of the evidence for this creation pointing to a much older universe. So what was Moses describing when he wrote about the creation of the universe? After all, he wasn't there when it began and was instead noting what God relayed to him about His creation. God didn't need to have Moses understand or convey an endless cosmos, the structure and function of atoms, that stars are actually suns, or that the band of dense stars in the sky are actually part of a rotating galaxy. The more important question for us to ponder is, what did God want Moses to convey? It does seem that God wanted to make it clear to Moses and the world that He, God, created everything, starting with an endless cosmos filled with incredibly designed and diffuse matter. And it was God who watched over His creation to form everything in it, no matter how long it took.

We also know that God is timeless, and to Him, He would not need to be in a hurry to do anything. Could the thinking about a young universe be yet another example of mankind taking what God has given them, then trying to fit it into a concept where mankind is more significant? So why should we adopt a limited time frame that matches mankind's written history on this planet? Are we again putting ourselves in the center of our universe?

In viewing both of these options, we can also conclude that Moses was correct in his dissemination about the sequence of creation. Everything fits in order. Matter had to first be created in an infinite space. This matter would need to be drawn into celestial bodies, and sunlight would eventually form when stars became dense enough to self-ignite. The only question is, how long did it take to create the cosmos?

However, I'll concede that if God wanted to make the cosmos appear to be very old, when it really isn't, He could have created it complete with everything viewable from across the universe and set into motion. After all, He created and distributed, on a mass scale, everything within this existence. But the big question is why would God be in a big hurry to create the cosmos? After all, He's timeless.

This debate should not cause a stumbling block for believers or cause us to think the Bible is in error. Moses' brief description centered more on providing the foundation that God created. Whether His creation occurred during a brief period or over a long time span, remember God could have made either occur. But if we are able to provide a synergy between God's Word about this creation and scientific observations, maybe mankind will finally conclude that the Old Earth option proves the existence of our Creator, who continues to interact within His creation.

Chapter Thirteen

Waters of the Deep

How really deep is that water?

THERE IS ANOTHER clue to this mystery in the second verse of Genesis 1. It notes that the Spirit of God was moving over the surface of the waters. Growing up, I assumed this described God's Spirit as moving over the oceans here on Earth. I could picture Him floating about in the atmosphere above a darkened Earth, recalling that God had not yet created sunlight. But if the Earth was without form, then what are the waters Moses described? Could these waters in fact be found in outer space instead of just being on the surface of Earth?

If you remember our discussions in Chapter 2 of this book, you will recall that during Moses' time, the popular view of the cosmos was based on a geocentric universe. During that time, the common belief was that the oceans on Earth and the blue sky above were both oceans of water. The Earth floated on a vast ocean of water, and the celestial bodies were suspended in the sky between us and the waters of the deep ocean that hung overhead. Was this what Moses was trying to describe in his reference to waters? If so, this does not mean that the Bible is in error. It means simply that Moses was describing the creation of the cosmos in terms that could be understood by people of his time.

As we can see, the creation account of Moses goes on to describe the sun being created with the separation of daytime and nighttime. He goes on to explain in the Genesis 1:6-8, that on the second day of creation, *"God said, "Let there be an expanse in the midst of the waters, and let it separate the waters from the waters.". And God made the expanse, and separated the waters which were below the expanse from the waters which were above the expanse; and it was so. And God called the expanse heaven."* Is this another reference to relating information relative to the common view about a geocentric cosmos?

Many in the religious community have assumed differently that the separation of the waters was in fact a separation of physical waters, where the oceans and lakes were separated from a thick cloud layer of moisture in our atmosphere that moistened the Earth with dew and didn't produce rain until the world flooding during Noah's time. But if we are to believe this interpretation about the waters, how would we explain the overwhelming evidence that water has flowed throughout the history of the Earth, causing erosion to shape the world into what we see today? God doesn't need the skies to be filled with a dense cloud layer, waiting patiently to form the first rainstorms as part of the great flood on Earth. If God can create everything, He can also create water to flood the Earth when He wants to, and then take it away to reestablish dry land. Besides, if there was a thick cloud layer above the surface of the Earth, how would they have been able to view the sun, moon, and all of the heavenly bodies?

Also, if there had been no rain until Noah's time, how would there have been any freshwater rivers and lakes? Where would they get their water? I know there are several examples of what appears to be freshwater upwelling throughout the globe, and several of these are found in the Holy Land. But water doesn't flow uphill against gravity. Each of these examples are fed from subterranean mountain sources that receive their moisture from rain and snowfall accumulations that soak into underground aquifers, only to emerge at lower elevations.

There is another reference to this same description, found in Proverbs 3:19-20, which states, *"The Lord by wisdom founded the earth; By understanding He established the heavens. By His knowledge the deeps were broken up, And the skies drip with dew."* In this reference, which was written by Solomon roughly 900 years BC, we can see a similar reference about the separation of the waters. During this period, a geocentric universe was still being postulated as defining the universe. What's interesting in this scripture by Solomon is his reference, in the present tense, that *"the skies drip with dew"*. This passage in Proverbs was referencing statements made by Moses about creation, but it also infers that during Solomon's time, which was after the great flood, that rain was described as dew. Could the reference to "drip with dew" be referring to rain? Could Moses have also described rain as dew? If this were the case, then the separation of the waters could in fact be a description of the geocentric universe.

While we're on the subject of the great flood, another concern I've seen expressed about this event has been how Noah could collect and redistribute animals and birds from other continents. Surely the kangaroos didn't hop over from Australia? What will really bake your noodle is the possibility that when God flooded the world during Noah's time, that this may not have been a total global flooding. It may have only flooded or totally submerged a portion of the globe which represented the area of the biblical world? After all, when God parted the Red Sea and made the sea floor as dry land for the escape of the Jewish nation under Moses, it was evident that God could confine water to a specific location until He chose to reflood the area. God could also have done this in reverse, by flooding a segment of the globe until He decided to dry up this same area.

Some have also theorized that the flooding occurred as the Mediterranean Sea reflooded the Black Sea, which may have been dried up prior to the time of Noah during the last ice age. They derive this possibility from evidence of recent discoveries of past human civilizations at the bottom of what is now the Black Sea.

We could also imagine that if God wanted to flood the entire globe but wanted to save all the animal species during this great flood, he could have transported all these creatures, two by two, to the ark and returned them miraculously after the water subsided. As to which possibility is correct, God only knows.

To summarize this chapter, it makes more logical sense that Moses' description of the "Waters of the Deep" in the first part of Genesis is simply noting that God created what was visualized in the heavens and on Earth by reciting it in terms that would be understood by the people at that time who spoke of the cosmos in terms of a geocentric concept. After all, if Moses had been enlightened to describe the cosmos in terms of atomic structure, and that the heavens were filled with galaxies containing billions of suns, no one would have believed him. The important reminder is to note that God created the heavens and the Earth.

Chapter Fourteen

Is Creation Possible?

*If it can be accomplished on a small scale,
why not on a grand scale?*

I'VE ALWAYS MARVELED at the stories told in the gospels about the miraculous healings performed by Jesus Christ. Not only are there a variety of specific miracles described, but there are also testimonials that Jesus Christ would heal everyone within a specific town. So it's safe to say that He performed tens of thousands of healings. The records also state that only a small fraction of these miracles are noted in these gospels. The specific healings noted in these books of the Bible leave us with evidence that God can heal every type of abnormality, even bringing several people back from the dead. So how did Christ perform these healings? Did he work at the atomic or subatomic level? Did he defy the laws of physics? Let's explore these questions.

It might seem to be a simple notion that someone would be healed of a deformed pair of legs, as an example. We would just claim it as a miracle and move on. But how would this healing actually be accomplished? The leg bones would need to be stretched out and possibly lengthened. The muscle tissue, ligaments, and tendons would all need to be built up and strengthened. Increased blood flow would be required. The tendons and nerves would all need to be repaired, and shin tissue would need to be extended to cover the longer limb. The mind would have to be altered to develop the memory and reflex responses needed to coordinate the movement of these rebuilt limbs, which would enable the person to rise up and begin to walk. Lastly, this would need to be done without the recipient feeling any pain. It's a lot more complicated when you think about everything involved in bringing about this healing.

Let's look at another example where Lazarus was raised from the dead after four days. In four days, his body would have been seriously decayed. All of the cells in his body would have been destroyed, and his memories erased. Yet Jesus Christ was able to restore him to life with a shout.

In these examples, as evidenced in the Bible, Jesus Christ was able to rearrange matter, energy, and restore memories in an instant. This rearrangement had to have been accomplished at the most minute level, which would mean that this transformation took place at the atomic or subatomic level. In other words, Christ knew every atom and brain memory in the person's body and orchestrated the flawless change. This transformation would have also included a neurological response that blocked any sensation of what was taking place. Truly miraculous.

There's one other miracle I'd like to illustrate that gives us evidence of God's ability to create. Do you recall the first recorded miracle where Jesus Christ was implored by his mother to change water into wine? Mary had asked her son to provide more wine at a wedding celebration, which He could have done by merely filling empty vases with newly created wine, or He could have transported wine miraculously into these vases. No one would have noticed the creation of this wine. Instead, Jesus asked the servants to first fill empty vases with water, which He then transformed into wine.

Did you ever wonder why the vases were first filled with water? I believe this leaves us with a clue about God's ability to create. To convert water into wine, Jesus would either have needed to first remove the water and then place wine in the vases, or he would have needed to transform water into the complex chemical composition of wine. The Bible does state that the water was transformed, which does make more sense, because why would water be placed in the vases only to be removed and replaced by wine? But to convert water, which is made up of only hydrogen and oxygen atoms, into a mixture that included carbon and organic compounds, would mean that Jesus changed the makeup of atoms within the mixture without causing any adverse reaction.

Let me explain why this is so miraculous. If we tried to fuse lighter-weight atoms like hydrogen and oxygen into carbon in the quantities needed for the party, we would have caused a fusion reaction that would have exploded and destroyed the whole region. When we cause atoms to fuse together, we see the release of large quantities of energy, but not so when God creates and transforms. This is evidence that God, through Jesus Christ, could defy the laws of physics and create without any adverse reaction. By the way, everyone at the party thought the new wine was really good, even though it hadn't been aged.

Knowing that God demonstrated his ability to miraculously create on a small scale, how could a universe-wide creation take place? So you tell me—which is more difficult to perform? Healing thousands of people with medical abnormalities, and raising people back to life who were dead (while also restoring their brain functions and memories), or creating perhaps an infinite amount of matter within the cosmos? I would argue that both are beyond our cognizant ability to understand, but we should conclude that God has the ability to perform on a large scale what He has performed on a small but complex scale.

We need to remember mankind's failed record of trying to limit God to a concept that fits within his ability to comprehend. Therefore, if the infinite God demonstrated His creative abilities to mankind when His son walked the earth, He also has the ability to create on a much grander scale that is beyond our comprehension. That might also include His creation of everything in the cosmos, with a perfect and identical design, instantaneously.

We also need to remember that the evidence is overwhelming that the cosmos had a beginning, and that it wasn't formed from a reoccurring Big Bang. The only solid evidence we have of the ability to create is that given by God and demonstrated through His son, Jesus Christ. Logically, we should also conclude that God was therefore the one and only Creator of the cosmos.

Chapter Fifteen

How Far Back Does Creation Go?

How much time was needed to create what we see today?

As we have elaborated in previous chapters, we know that the universe will end when gravity and time consolidate everything into massive black holes, leaving the heavens dark, cold, and void of any visible matter. This has led us to the realization that there had to also be a beginning of everything. Since we have successfully ruled out the flawed Big Bang theory as a beginning, we can now conclude that this beginning would properly be characterized as the Great Creation. But how long ago did this occur? Was it instantaneous throughout the cosmos? Let's discuss the possibilities.

It's true that God could have started the creation at any single point in time. But if God is timeless, why would He have been compelled to create it a mere 6,000 years ago? It seems more logical that His creation began a long time ago, which gives us the visual and gravitational evidence of a very old cosmos. The evidence of light travel alone shows us that photons have been moving through space for tens of billions of years.

Of course, God could have created the heavens, as well as photons of light, already in their trajectory just a few light years from Earth. But if we travel throughout space, will we still find the existence of light photons emanating from distant stars? If we found these light photons throughout space, then the logical explanation is that these distant objects are there and that the cosmos is truly very old. We have been able to calculate the location of distant galaxies by measuring the diameter of supernova explosions that occur within these galaxies. If God created these distant galaxies, He would need to make sure that light photons traveling from these galaxies provided the necessary information to identify their relative position in the universe. But why would God be compelled to create a young cosmos with all the evidence that it is very old? It seems conclusive and more probable that we are a part of an ancient cosmos. So let's see if we can get a sense of when it began.

If, in the beginning, God did create the heavens and the earth and the earth was without form, just what did this ancient creation beginning look like? Did this beginning involve either the instantaneous creation of all matter, or the creation of all matter occurring over a long period of time? Whichever the case, just think about how wonderfully powerful God is to have developed a perfect design for everything, and then spread His creation into an infinite cosmos. Let's look at some possibilities for how this Old Earth creation may have started, and how long ago this may have begun.

If the creation began with everything being placed in the cosmos, as stated in Genesis, and there was no form to its existence, could this creation have only included the formation of hydrogen? It would be without form, invisible, and it would contain the necessary building blocks for all matter in the universe. As previously stated, over an extended period, hydrogen by itself could condense into dense clouds of liquid hydrogen. But is it possible that these clouds of hydrogen gas could be dense enough to ignite into a stable fusion reaction that will sustain a sun without blowing itself out?

I doubt that a sun made up solely of hydrogen could become stable, as a sun needs a dense gravitational field to contain a stable fusion reaction. I know that for years scientists believed that our sun was solely composed of hydrogen gas, but the fusion reaction alone combines hydrogen into helium and possibly heavier elements, so we know other matter exists in our sun.

We also know that other objects in space collide with our sun. So it is only logical that our sun has a dense and solid core made up of heavier elements. These elements provide the gravitational forces needed to contain the sun's fusion reaction. We also know that the sun contains vast amounts of iron, which form the basis for the sun's ever-changing magnetic fields.

But let's say that the creation of hydrogen alone could have formed the basis for the beginning. Then how old would our cosmos be? It would be incredibly old, as it would take trillions of years just to have fused simple hydrogen into all of the different elements, let alone the time then needed to form all of the celestial bodies we see in space. After all, the heavier elements would most likely have formed as part of the collapse and destruction of suns in supernova explosions. This would need to be repeated time and time again to create enough heavier elements to form planets.

However, I believe it is wonderful to think that God could have begun the universe with such a simple beginning, knowing that through His laws of physics, all the elements of matter would form through fusion reactions. This would eventually lead to the creation of the heavens and our miraculous world.

A second possibility could envision that God created all the elements as part of the first day of creation, but that this material was so scattered it had no form. Over time, as the Spirit of God moved over His creation, He could have caused this matter to form into celestial bodies and allowed the forces of gravity to combine, shape, and mold the cosmos. This could have occurred in less time than the first option we discussed, depending on how active God was in moving and shaping His creation. But I would note that even in this case, the time needed for gravity to pull everything into planets, suns, and galaxies would take much longer than the tens of billions of years prescribed by those trying to force fit the Big Bang creation theory.

There are probably a countless number of options about how creation could have started, but for simplicity's sake, we'll just discuss one other option. In this third option, God could have started by creating individual hydrogen atoms and then molding His creation by supernaturally fusing hydrogen into heavier elements as He was hovering over them. It's an interesting concept, and this could have taken some time to complete, depending on how involved God was at moving His creation along.

We can conclude the following from this discussion. The first is that the creation of the cosmos is most likely much older than anyone has been postulating. Secondly, when we consider that gravitational forces and the fusion of matter, along with God's creative talent, can form everything from giant spiral galaxies to a delicate beautiful butterfly, we see that the cosmos has a perfect design. Thirdly, the timeless God of creation has been at work hovering over His creation, taking His time to form what we see today.

Finally, unless the religious community can provide evidence supporting the New Earth theory, they should let go of this outdated perspective about creation and realize that a long-term creation fits within Genesis 1. For believers to continue to cling to their narrow perspective that the first day of creation was only centuries ago, puts them at risk of being ridiculed by non-believers who view these perspectives as out of touch with the evidence around them, not unlike the members of The Flat Earth Society.

Chapter Sixteen

Creation vs. Evolution

Are evolutionists just monkeying around with creation?

KNOWING THAT THERE was a beginning to the cosmos and that this creation was not the Big Bang, can we assume the forces of gravity and the influence of evolution alone were responsible for the continuation of the creation story? The other side of this question ponders whether God, after making His creation, stayed involved in its progression by continuing to intervene and mold the cosmos and everything in it over time. So, which is it? If God was hovering over His creation at the beginning, why wouldn't He still be involved? Could all of the changes in species have been caused by God, in His timing, to evolve and change this planet to support what He wanted to see in His creation?

Ever since Darwin theorized the evolution of everything, there has been a constant debate as to which came first, the chicken or the egg. This debate still rages today, but we see it expressed more by the scientific community pursuing discoveries as to how biological systems function, rather than exploring how it is that these complex systems are able to replicate without random mutations causing the decline in the biological functions within a given species. Ah, but there is another buzzword used by evolutionists, which theorizes that "natural selection" will cause the survival of the superior mutated species. But other than species getting larger or smaller or cross-breeding with other animals on rare occasions (which have mostly been manipulated by mankind's intervention), we don't see evidence of positive evolutionary changes taking place.

Another point to make about evolution through mutation is that this process needs to be sustainable. The mutated or changed creature would need to replicate. Cross-breeding typically leaves the new creature sterile, like in the case of a jackass, which is bred from a mule and a horse. So in the chicken and egg scenario, for a species transformation to be successful, the chicken would need to either produce multiple identically mutated new creatures that could then successfully breed to sustain the existence of the superior creature, or the new mutated creature would need to have dominant DNA that would foster the continuation of the new species when it mated with a non-mutated clone of its parent.

I believe this riddle of evolution versus creation can best be resolved by looking at this subject from a microbiological perspective. Remember Darwin's warning about the complexities of biological science. For starters, just look at the human body with all its intertwined and complicated systems and try to convince me that this evolved. Let's look at eyes, for example. How could the mutation of light-sensitive cells ever evolve to formulate the complex design and function of eyes that we see on almost all living creatures? Let's list a few of the complexities involved in eyesight:

- Two eyes for depth perception.

- Articulation of eyes.

- Ability to focus near and far based on lens articulation.

- Placement of eyes in a moistened eye socket fed by tear ducts.

- Eyelids which could open and shut to keep the eyes moist.

- The retina, able to sense light photons across a wide spectrum of light.

- Circulation of a clear fluid in the eyeball to keep a clear separation between the lens and the retina.

- Blood flow to the eyes from a tethered blood vessel that still allows the eyes to rotate.

- Contracting iris that responds to variations of light intensity.

- A complex muscle system that enables substantial rotation of both eyes in unison.

- A neurological response system that translates captured light on the retina to a clear visual image in your mind.

I'm sure you can think of some additional unique design features contained in your eyes. You can apply this same review to other complex systems within the human body and conclude that it truly is miraculous.

I think, however, that the best example we can use to illustrate the complexities of life centers around the amazing process in the duplication of cells within all living creatures. For each cell to replicate, it needs to first divide its DNA into two separate strands. This complicated process involves several different protein-based enzymes that act like a zipper to first separate the long DNA strand. Other enzymes come along and duplicate the missing sequences in the DNA strands, which helps to form two identical strands within the cell. Once this is accomplished, the cell separates internally and divides itself into two identical cells. There is no way this cell division is a product of natural selection. It's too complex to have begun long ago due to a random selection process. As stated before, Darwin believed that if future discoveries in biologic science occurred which added even more complexity than existed during Darwin's time, then his theory of evolution could not be supported.

Another question about cell division is how do cells know how to create a living creature? Yes, the DNA strand has all the information about the creature it is attempting to generate. But you would think that cell division, which is replicating identical cells, would result in a blob of identical cells, not a human being. So how do cells know how to create a complex body? Yes, the code to form a body is contained within the cell's DNA, but how do individual cells know where they are in this newly forming body and how do they know and decide what their cell division will form? Is this simply a complexity we have yet to understand, or does God intercede in the formation of all life?

Also of interest is why our bodies produce fewer cells as we age, which limits our life span. What triggers this decline in cell production? Scientists have been trying to understand this phenomenon to see if they can reverse or retard this decline and extend our lives. For me, the answer lies in the belief that the Designer affixed a limited span of life to humble us to an appreciation of what we have, while we have it. Remember, nothing lasts forever in this reality, not even the earth and sky.

Have you ever wondered why there are so many different plants, animals, and insects on this planet? The design of life is so infectious on this planet that it's everywhere and in every form imaginable. When we look for life elsewhere in the cosmos, we would probably be amazed to find a few microbes on some distant planet, because random selection will never duplicate what we have here on Earth. You should also drop your perceptions, fostered by *Star Trek* and *Star Wars,* that complex life forms exist throughout the universe, as I truly believe this world is an exception.

It's also interesting to note that, so far, we have found that other planets and asteroids are composed of rock and dust, and we don't see any separation of minerals on these bodies like what we see on Earth. Have you ever wondered why this planet has pockets of mineral-rich compounds that allow them to be mined and extracted, which gives us various materials that we use in our daily lives? Why does one mountain contain the elements of copper, or silver, and another location contains the elements needed to produce concrete? I do believe that plate tectonics can explain why we might find deposits of hydrocarbons and gold-bearing rock. But why would we see the separation of minerals and not more of a homogenous mixture of elements like we see on the Moon, Mars, or Venus? This may very well be the product of progressive creation, which we'll explore in the next chapter.

    I would conclude from this chapter's discussion that the probability of the existence of Earth and all of its life forms are not a result of evolution, but the overwhelming evidence of a marvelous creation.

Chapter Seventeen

The Progressive Creation of Earth

What are the odds of us getting here through random actions?

WE HAVE BRIEFLY discussed the concept of progressive creation and how this may have played a role in the creation story. To recap this theory, we would define progressive creation as one where God continued to hover over His initial creation, shaping and molding it over time into what we currently see today by supernaturally intervening in the natural progression of events. This progressive creation would apply to anything or anywhere God chose to advance His creation by causing miraculous changes to occur. Progressive creation could have shaped the heavens and been responsible for miraculous changes that have helped mold the earth and all its abundant life. You may ask, what's so miraculous about Earth? After all, we are here now, and everything exists around us that's needed for us to flourish. We might also ask, why wouldn't this pattern of life be duplicated on other worlds?

Let's first look at what would be needed to create an Earth-like planet with all of its life forms. You'll find that when we look at the large number of events that would have been needed to take place in order to form what we see around us, the probability of this occurring through random coincidence is utterly impossible. Scientists have looked at the probability of there being other Earth-like planets within the Milky Way galaxy that could duplicate what would be needed to sustain an environment suitable for life as we know it. They have concluded that there might be only nine planets in our galaxy that could be suitable for life. This is an extremely small percentage, considering our galaxy is made up of billions of stars. When you then look at what would be needed to create, sustain, and evolve complex life on those nine worlds, you could easily conclude that sustainable life within our galaxy is highly improbable. The only possible explanation for how this planet may have evolved would be by the intervention of God to create and sustain His creation through miraculous changes. Here are a few items needed to duplicate an Earth-like planet that could sustain abundant and evolved life.

- An Earth-like planet would need to be located in a solar system with a stable and relatively small sun, which would be separated and far apart from other stars. This would prevent frequent cataclysmic events from adjacent stars impacting Earth's development, like when a star collapses in a supernova explosion.

- You would need sufficient separation of this planet from other celestial bodies to prevent more frequent cataclysmic collisions.

- You would need much larger planets in your solar system, like Jupiter and Saturn, which are located farther away from the sun than the Earth-like planet, to reduce the number of asteroids that could impact the planet. These large planets tend to draw off many of the asteroids that would otherwise impact Earth.

- When Earth was formed through the process of accumulation of asteroids, there needed to be the correct mixture of elements to enable the development of a solid iron core, which could stay heated for billions of years. This was most likely achieved by the inclusion of sufficient fissionable material deep within our planet to fuel the continuation of thermal activity. This prolonged thermal process has allowed the distillation and separation of different elements within a relatively thin crust.

- The planet would need to be rotating at a similar rate as Earth so that planetary temperatures wouldn't become extreme.

- A rotating iron core would be needed to produce a strong magnetic field like we have on Earth, which prevents the sun's solar winds from stripping away the atmosphere.

- The planet would need to be located within the "Goldilocks" zone in its orbit around the sun. Not too close to the Sun, which would burn off its oceans, and not too far away where the oceans would freeze solid. Both extremes would preclude the development and sustainability of life.

- Size matters. If an Earth-like planet were too large or small, it would most likely either be a gas giant or a cold stone in space.

- The capture of a large moon would be necessary to enable the Earth-like planet to have tidal fluctuations, which help to cleanse the ocean shorelines from stagnant conditions.

- Sufficient solar winds are needed to drive atmospheric winds on an Earth-like planet, which brings about atmospheric fluctuations needed to produce rainfall across the surface of the planet.

- An Earth-like planet would need to rotate in a stable but tilted rotation that would expose its entire surface to solar heating. Otherwise, unexposed sections of the world would freeze over, causing a permanent planetary ice age.

- To duplicate the composition of Earth, you would need a similar mixture of elements.

- Plate tectonic action would be needed to mold and shape the dry land. These massive plates float on a molten mantel deep within the Earth. Thermal activity within the Earth causes these plates to slowly move. If these plates had not formed, then erosion would have eventually caused the continents to erode into the seas, and the Earth's entire surface would be engulfed by water.

- Plate tectonic subduction zones are needed on an Earth-like planet to foster the production of hydrocarbons by submerging carbon laden material deep below the Earth's surface where it can be heated to separate lighter elements. Over time, the lighter-weight carbon will eventually gravitate upward through the planet's crust, where it can be trapped under dense rock formations.

- You would need a planet to form a crust on its surface through thermal interactions which occur deep within the planet's subterranean mantels. This thermal process distills different elements, like gasses, water, and carbon, which are necessary building blocks for carbon-based life.

- With the churning of an Earth-like planet's crust over time, due to tectonic action, as well as with erosion and corrosion forces at play, I find it very peculiar and most improbable that we find anything but a homogeneous mixture of elements within our earth's crust. It's true that we mostly find that Earth's crust is made up of rock formations, from massive granite mountains to arid desert sands which were formed by the forces of erosion and corrosion that break down rocks to their smallest forms. But why do we find specific deposits of separated minerals on Earth? In one location, you find a deposit of copper, or silver, or iron, which can be extracted and refined. This seems to be highly improbable to have occurred naturally. So let's conclude that it would be difficult to find similarly separated naturally occurring mineral deposits on other worlds.

- You would need a planet with a high concentration of water to begin and contain the building blocks of life, and to have sufficient water for sustaining fresh water sources through abundant rainfalls. As an interesting side note, the design of water molecules enables it to expand when it freezes. Something we all know. But most liquid elements shrink when they freeze. Why is this important? Well, if water became denser when it froze, then ice would sink, causing the oceans and lakes to freeze solid from the bottom up, which would cause Earth to be a frozen planet.

- Another essential element to the formation of an Earth-like planet is the salinity of its oceans. The vast quantities of salt found in our oceans prevent these bodies of water from becoming stagnant. Could you imagine what it would look like if the oceans were not salty? They would be very slimy, and stinky, causing the lack of oxygen to choke out the development of life. The oceans on Earth are the great filters of this planet that cleanse vast quantities of contaminants that wash into them, only to deposit this material on their ocean floors where they can be slowly drawn under the continental tectonic plates over millions of years.

- It's also important that if you were to duplicate Earth, you would need to form a planet that has minimal concentrations of toxic elements. There could not be any high concentrations of arsenic, lead, and other harmful elements to interfere with carbon-based life. It's a good thing our oceans are filled with a concentration of salt, instead of arsenic. Otherwise, we'd have no life at all in our oceans.

- Also needed is an atmosphere that allows for the evaporation and condensation of water to foster life on land.

- Early waterborne life on Earth converted the atmosphere from mostly carbon dioxide to an oxygen/nitrogen mixture. A partial oxygen atmosphere could not have formed without this interaction, but God could have also been involved in this causality within His progressive creation.

- With the conversion of our atmosphere to an oxygen-based one that sustains life, an inert gas was needed to prevent an explosive mixture. Nitrogen fulfills this role and does not adversely impact carbon-based life. It's a good thing helium didn't replace the role of nitrogen or we'd all be singing the chipmunk Christmas song.

From this discussion, one can see how unique the formation of planet Earth is, which begs the question of how could this have occurred on its own? Could Earth have formed as a result of random collisions in space, or was it formed and transformed through God's intervention over time?

So far, we have only looked at one side of the equation about the formation of Earth. Let's now look at what would have been needed to develop, sustain, and evolve life on Earth. It's easy to look around and see all the abundant life and conclude that it must have just started somehow, and most likely in a very simple form. But that "somehow" is a little more complex than could occur through random or natural selection. Also, the transformation of simple life to complex beings was no accident. Let's look at some of the requirements for life to have begun on Earth and then evolved to its current condition.

- For life on earth to have begun, it most likely began from a single-celled creature. But the more we learn about the complexities of even those creatures, we see that its design is far more intricate than we imagined. How could these creatures initially be formed with a DNA structure and the ability to replicate? The intricate process involved in splitting DNA strands within a cell is so complex there is no way this could have evolved from nothing.

- How could single-celled creatures have mutated over time to eventually form creatures with millions of DNA sequences while maintaining the ability to replicate?

- How do cells within living creatures know which cell to form when they divide? There are all different types of cells and entirely different types of structures within living creatures. Cells can't just be replicating identical cells. If this occurred, we would be nothing but a blob of cells.

- Why is there diversity of life, but an identical structural design? Just look at the similarities between all of the animal, bird, reptile, and fish species on Earth. You would think that if random mutations and natural selection were the only causes for change, we would see more randomly structured creatures.

- Why were there sudden extinctions, which were most likely caused by asteroid impacts, followed by rapid changes in animal species? Was this God's way of saying your reign is done, Mr. Dinosaur, and now we're moving on to the next chapter in My creation story?

- Why are there still monkeys if evolutionists believe we evolved FROM monkeys?

- Why are flowers and butterflies so colorful when many of the creatures that interact with them are colorblind?

It should be pointed out that progressive creation does not mean that life evolved on its own once it got started. To the contrary. When God decided it was time for man, He then created man, using the science and design features of other creatures He had created. Why would God have chosen a totally different design, when He had already perfected a resilient pattern of life?

So, getting back to the chicken or the egg debate, progressive creation falls on the side of the chicken, which was created, male and female, with the ability to survive and have offspring. Once created, their environment played a role in their progression as a species, i.e. bigger, plumper, juicier, better tasting, so everything could taste like chicken.

Conclusions

To summarize this chapter, we can conclude that the probability of Earth's formation—with all its abundant life—absent God's hand in His creation story, is highly unlikely. But some skeptics argue that the formation of life has a perfect design and therefore must have occurred with the help from some intelligent life form. But, if so, then who created that life form, which would have had the same impossible odds of ever forming on its own? It's obvious that the conclusion can only be that there is only one Creator of life in the cosmos, and I thank God for the opportunity to live amongst all His creation.

Even though it's next to impossible to think that similar worlds with all sorts of abundant life like Earth could exist in the cosmos, I would be remiss if I didn't postulate about the possibility that God may have chosen to create other populated worlds. He could be interceding in these created worlds as well. After all, why are we so special to think we are the only divinely populated planet in the entire cosmos? Remember our review of history has shown us that mankind tends to believe that everything revolves around our existence. The Bible doesn't mention this possibility and doesn't need to. After all, God has chosen to create this planet and has shown His love and mercy to us, independent of anything else that may be out there in the cosmos. If life on other worlds does exist, it would not come as a surprise to God, as he would have hovered over that world as well.

We should also remember that solar systems within our region of the Milky Way galaxy are extremely spread apart. There is so much debris in space one could run into while attempting to travel great distances in a short period of time, that it would be next to impossible to safely travel to other worlds outside our solar system. Deflector shields sound like a great solution for deflecting objects in space, but you would be traveling too fast to ever be able to detect objects in your path before you ran into them. After all, a rock the size of a pea could destroy your spaceship that's traveling at velocities approaching the speed of light. This vast separation in space may be God's way of keeping other worlds from interacting within His creation, if these worlds even exist. So, no Star-Trekking to other abundant class M planets!

Chapter Eighteen

The Link Between Science and Religion

*God created science, and man has
been trying to define His work.*

THROUGHOUT OUR ADULT lives, we've been bombarded with different conjectures about creation and evolution. We rarely delve into the details of this debate, but simply note the differing opinions which have caused a tug-of-war between science and religion. It's fair to say that both sides are so deeply entrenched in their beliefs, while accusing each other that they can't possibly be correct in their assessments. One of the main reasons I've been inspired to write this book has been my desire to show a link between scientific observations and God's creation. We'll see if we can provide sufficient evidence of this link and find a common ground that brings these two sides together without causing either side to relinquish the foundation of their belief.

In this book, we have provided sufficient evidence to prove the following:

- Throughout history, mankind has been assumed to be at the center of his visible universe. As we have increased our ability to gaze further into space, we've expanded the borders of our known universe, but still find that our world is perceived to be at a central point of a limited universe.

- As mankind peers further into the cosmos, he increases his knowledge of its formation and finds that previous assumptions about the cosmos may have been in error.

- Gravitational forces can indeed slow down light photons traversing the universe. Therefore, the universe may not be expanding but slowly coalescing.

- It appears that the Big Bang theory was derived using an incorrect assumption about light travel. There does not seem to be any retrospective analysis of this critical assumption, which formed the basis for this theory.

- There is scientific agreement (and we have supported this argument) that the cosmos is coalescing slowly over time and will eventually end when everything is consumed by massive black holes. We can therefore assume that since the cosmos will end, that it must also have had a beginning and was not always in existence. With the Big Bang theory disproven as the beginning of the cosmos, the only possible conclusion is that the cosmos was created.

- Theories about the cosmos abound with no evidence to back up their assumptions. Some of these new theories are attempting to offer explanations for why previous theories, like the Big Bang, are found to be unrealistic. But in doing so, these new theories only compound previous errors. As an example, remember our discussion about dark matter and dark energy, which is being touted as filling the massive voids in the universe when there is no evidence supporting this theory. It also seems that others are just attempting to find notoriety by postulating new theories and touting their own supposed brilliance.

- There is overwhelming evidence that the cosmos had a creation beginning. The only disagreement centers on when this occurred. However, I believe we can conclude that for science and religion to find a common understanding about the cosmos, both sides would need to accept that the cosmos is much older than currently envisioned. The evidence presented in this book leads to this conclusion.

- If there is no divine intervention into our future, the universe will have a dark and cold ending when everything is absorbed by massive and dense black holes.

- Creation of the cosmos involved the outpouring of extraordinarily designed matter that is universal throughout creation.

- The existence of Earth, with all of its abundant life, is improbable without God's continual intervention as part of a progressive creation. Random natural selection could not account for the initiation of life and its ability to replicate. Also, complex life forms could never have evolved through mutations, which supports the notion that progressive creation intervention was involved to develop the world we see around us.

Conclusions

We can see from this list that there is a linkage between science and religion when you correct the deficiencies within the Big Bang theory and accept that the only logical explanation for our existence is a divine intervention with the formation of the cosmos, which includes the creation and development of Earth and all its life forms. This intervention would need to have occurred over time to bring about the miraculous changes needed to fully develop this planet with all of its advanced life forms. There is no need to overcomplicate solutions. Sometimes the simplest and most fact-based solution is the most accurate and best solution.

I've found that this exercise of investigating the origins of the cosmos has been like trying to describe an ocean, when all you see is a drop of water. But if you fail to first understand the drop of water, how can you begin to describe the sea?

Now That You Know

Now that you know the preponderance of evidence points to the Genesis Creation of the universe, how will this impact you?

My advice to the scientific community would be to stop digging yourselves deeper into abstract theories without supportive evidence. I would also note that from our historical review and from my practical work experience, all theories and assumptions should be checked and rechecked. There's nothing more embarrassing than finding out you've correctly solved an equation but initially failed to define the problem properly.

For those in the scientific community who have tried to find a nexus between science and religion in the creation of the cosmos, I hope this book has provided you with sufficient evidence to make this connection. My hope is that you take this work further to strengthen the concepts I've presented and for you to openly renounce the flawed Big Bang theory.

There are some in the scientific community that have attempted to disprove the existence of a Divine Creator. For whatever reason, they refuse to accept that the overwhelming evidence points to a divine creation. My suggestion to you is that you should consider all of the evidence based on facts and open your mind to the possibility that the only solution to the mystery of the cosmos points to the existence and involvement of God. There is no other explanation. Knowing this, enlarge your perspective and appreciate the wonder of this creation design and accept Him for what He is. Denying His existence and hardening your heart to the truth only separates you from Him.

For those of you who wish to still cling to the six literal days of creation described by Moses, don't let this book rock your faith or dissuade you from your belief. If the Creator of everything wanted to do it in six days, He could have. However, this thinking still leaves open the question of why God would create in six days, when all the evidence of this creation supports it occurring a long time ago? If God created everything, as well as the science behind His perfect design, why would He override the laws behind His science to present a false image about the timing of creation? By limiting your perspective to a New Earth theory, you also appear to be close-minded to a possible link between the two extremes of science and religion.

For those of you who believe in a much older creation, this book will hopefully provide you with sufficient details to enable you to see the nexus between science and religion and confirm that, without the Creator, there would never be this wonderful world in which we live. I hope this book strengthens your faith, knowing more about God's creation and how it is backed up by scientific evidence.

It may seem like I've raised more questions in this book than I've solved. If nothing else, my desire has been for you to think about the connection between science and religion. So the next time you find yourself under a brilliant star-filled sky stretched from horizon to horizon, spend a few minutes gazing into the infinite heavens and think about the possibilities.

Chapter Nineteen

Theories Derived

IN THIS BOOK I've postulated and, hopefully, proven several theories. Some of these theories may have already been adopted as factual by some in the scientific community. I haven't completed an exhaustive review to see if others have come to these same conclusions, but I feel it is important to note what I believe to be different perspectives about the reality of physical science. These theories were covered in more detail earlier in this book. Here's a list of these theories with a brief description of each.

- **Matter and Energy**

    Einstein theorized that matter and energy are interchangeable. However, the energy released from nuclear reactions is not caused by the depletion of matter, as theorized by Einstein in his famous equation $E=mc^2$. Protons, neutrons, and electrons are neither destroyed nor created in the process of nuclear fission or fusion. The energy released is purely caused by the separation or combining of particles within an atomic nucleus. When fused or split, different elements produce different quantities of released energy, which is not proportional to the mass of the given element. Therefore, this famous equation is not proportional, and should be replaced with a simple statement that there is a significant energy release from atomic nuclei when they are either fused or split, which is proportional to the specific reaction of the given elements.

- **Eliminating the Speed of Light Restrictions**

   Einstein theorized that objects accelerating to the speed of light are slowed and prevented from exceeding that speed. This theory was derived from observations about sound waves, which have an entirely different mechanism behind their behavior. This theory could only occur if there were some interacting forces throughout space that would limit acceleration. As far as we know, this limiting force does not exist. In fact, recent Hubble telescope findings verify that objects can travel faster than the speed of light.

- **Particle Acceleration**

   We have theorized that all accelerated particles and forces begin their journeys at the same velocity, which is equal to the speed of light. They don't accelerate from rest and then slow their acceleration as their velocity reaches the speed of light.

- **Gravitational Effects on Fast-Moving Particles**

    Gravitational forces emanate from an object at the speed of gravity and are able to then influence the attraction of distant objects at the same speed of gravity. If particles are already traveling toward an object at velocities approaching the speed of gravity or light speed, the force of gravity from the attracting object will have minimal ability to accelerate the approaching particles.

- **Photon Deceleration**

    Photons of light leave an object instantaneously and begin their travel at the speed of light. However, the forces of gravity throughout the universe will eventually slow these particles and elongate their electromagnetic wavelength as they traverse the universe.

- **Coalescing Universe**

    The universe is not expanding or contracting from the effects of a singularity explosion. The universe is, however, coalescing over time. This will not lead to an eventual recreation of a singularity. When you take into account the evidence that light photons reduce their velocity over time, we can observe that all matter in the universe is staying within a stable orientation, only being coalesced by gravitational forces from adjacent celestial bodies.

- **Dark Matter and Energy Don't Exist**

    There is no evidence of their existence, and they were only conjectured to support the theory of an accelerated timeline under the Big Bang theory. The discovery of black holes has negated the need for this theory. Their existence would have distorted the visual image of distant objects.

- **Science Defines Creation**

    Science supports the Great Creation of the cosmos. After all, who created science? All of the physical evidence points to a massive creation of the cosmos with a perfect design.

- **Infinite Cosmos**

    The cosmos is infinite, but only God knows how widespread the existence of matter is. It is very possible celestial bodies exist throughout an infinite cosmos. Mankind just hasn't been able to see far enough into the cosmos to determine how far celestial bodies extend. The infinite cosmos and time itself are realities that continue to define our existence. Only the divine Creator has the ability to navigate and mold their future.

- **The Universe Is Extremely Old**

    All the physical evidence supports the theory that the universe was created and is extremely old. It is also much older than the Big Bang theory supposes.

Chapter Twenty

To Infinity and Beyond

Imagine the possibilities!

AFTER FINISHING THIS book, I thought it would be a worthwhile exercise to dream a little bit and explore some of the possibilities for realities we might find in an infinite cosmos. First of all, we need to understand the immenseness of an infinite cosmos. We have used terms suggesting the cosmos goes on forever, but can your mind really comprehend how big that is? Many of us assume that space goes on forever, but we're not sure if matter extends forever as well. For years I've concluded that if you can see far enough into the cosmos, you will find the end of matter or celestial bodies. Maybe this has been driven by our predisposition with the Big Bang theory. But if matter extends to infinity, how far is that and what does that look like?

In my college calculus class, I remember solving equations relative to the number infinity. During one class we were assigned a particular equation which sought the answer to the division of infinity by infinity. I was surprised at first to find that the answer wasn't one, as we all had been taught that no matter how big a number is, when you divide that number by itself the answer is always one. Instead, we learned that when you divide infinity by itself, the answer is infinity. Or in other words, infinity is so large that you could never divide it to be anything smaller than infinite, even if you divided it by itself. Let that sink in a minute...

Now let's think of this in terms of size comparisons. Picture yourself being able to travel through space so fast that you could travel across the entire known universe in a split second. If you then attempted to reach the edge of the infinite cosmos at that same speed, you would never be able to reach it, as it would take you an infinite number of years to do so.

So, let's look at empty space. How was this dimension created, or was it? Did it always exist, with the Great Creation of everything contained within it coming sometime later? If you were able to travel back in time before the creation of matter and if you found yourself in empty space, how would you be able to perceive you are in an infinite cosmos? There would be no reference to anything. You could move your spaceship all around, but you would have no idea where you were going or even if you had moved at all, except for the forces you would feel with acceleration. You could even make the argument that the dimension in which you found yourself doesn't exist, because you would have no evidence of its existence. So, did space only become a reality when matter was created, or did God perceive the need to create space in order to house His infinite cosmos? We simply don't know when or if the dimension of space was created or if it always existed, but it's interesting to ponder.

In this three-dimensional space we have postulated that God created everything within the cosmos. If matter extends forever in this reality, what could you find out there? Does the same design of matter exist throughout an infinite cosmos? If it does, we would find an infinite number of inhabited worlds, even if God chose to create life on other worlds at the same frequency He has in this known universe. Within this reality, we would find in the infinite cosmos an infinite number of inhabited worlds with infinite possibilities of life. We would then find that somewhere, in the distant ether of outer space, an identical Earth with an identical history, and with people identical to the ones we see around us today, all the while living identical lives and having identical thoughts to the ones we are having. They might even be reading this same book and wondering if this could really be happening. It would be a total mirror image of our reality here on Earth. This is a reality of infinite possibilities. But it should also be noted that this possibility is totally driven by what God intended for his creation within his infinite cosmos. He may have planned something entirely different.

But what if matter within the cosmos ended at some point? What would you find beyond that point? After all, God would not have been obligated to extend all matter within the universe as far as the dimension of space itself. Would you just find one vast universe within the cosmos surrounded by an eternity of empty space? Or could God be creating and recreating vast universes within the infinite cosmos for some purpose we can't comprehend?

Pondering this also allows us a glimpse of the magnificent awesomeness of the omnipotent, omnipresent God, who could have created beyond our comprehension. And just think of His unlimited ability to simultaneously interact with everything and everyone within His creation, even extending down to the atomic level, and hearing our every thought and prayer.

So if God has the ability to operate on this detailed level when creating everything, and when transforming matter, and when also interceding on our behalf, is He separate from this three-dimensional reality which allows Him the ability to enter in and interact with this reality at His choosing, or is His existence a part of this reality and somehow connected to all matter within the cosmos, which allows Him to be omnipresent?

Within the infinite horizon, only God knows!

Chapter Twenty-One

Final Thoughts

The decision is yours for the taking!

I HOPE THAT the time you have spent reading this book has proven to be a worthwhile endeavor and that you were able to absorb many of the key points. For a number of years, it has been my sincere desire to share my findings about the spectacular universe that surrounds us. It gives me great comfort and peace knowing that we have found a nexus between scientific facts and God being the creator of everything.

For those of you who have already accepted Jesus Christ as your Lord and Savior, I pray that this book will give you encouragement in knowing that God has been in control throughout the history of the cosmos and has never left us and never will!

For those of you who have not yet come into a personal relationship with God through His Son, Jesus Christ, you have now been presented with overwhelming evidence that the cosmos was created and transformed by an infinite God who desires a personal relationship with you! The question to you is, "What will you do about it?" We can either recognize God for who He is and accept His offer of eternal salvation through faith in Jesus Christ and His death on the cross, or we can face eternal separation from God. The decision is yours alone to make, and God is waiting patiently for you. Don't delay!

Glossary of Historical Figures

Anaxagoras: Anaxagoras was a pre-Socratic Greek philosopher in Athens who theorized that the sun and stars were fiery rocks and that the moon was an earthlike sphere illuminated by the sun; 500-428 BC

Aristotle: Aristotle was a Greek philosopher and polymath during the Classical period in Ancient Greece. Taught by Plato, he was the founder of the Lyceum, the Peripatetic school of philosophy who theorized that the Earth was a sphere; 384-322 BC

Bruno: Giordano Bruno was an Italian Dominican friar, philosopher, mathematician, poet, cosmological theorist who believed in an infinite universe; 1548-1600 AD

Copernicus: Nicolaus Copernicus was a Renaissance-era mathematician and astronomer who formulated a model of the universe that placed the sun rather than earth at its center. He also identified the inner and outer planets; 1473-1543 AD

Darwin: Charles Robert Darwin was an English naturalist, geologist, and biologist, best known for his contributions to the science of evolution; 1809-1882 AD

Einstein: Albert Einstein was a German born theoretical physicist widely acknowledged to be one of the greatest physicists of all time; 1879-1955 AD

Eratosthenes: Eratosthenes oh Cyrene was a Greek polymath: a mathematician, geographer, and astronomer. He was a man of learning, becoming the chief librarian of the library of Alexandria. He determined the world was round and calculated the diameter of Earth and the distances between the Earth, moon, and sun; 276-194 BC

Galileo: Galileo di Vincenzo Bonaiuti de' Galilei was an Italian astronomer, physicist, and engineer. Galileo has been called the father of observational astronomy and of modern physics; 1564-1642 AD

Herschel: Frederick William Herschel was a British astronomer who was the founder of sidereal astronomy for the systematic observation of stars and nebulae beyond our solar system. He also hypothesized that nebulae are composed of stars and developed a theory of stellar evolution; 1738-1822 AD

Higgs: Peter Ware Higgs is a British theoretical physicist emeritus professor in the University of Edinburgh and Nobel prize laureate for his work on the mass of subatomic particles; 1929-present

Hubble: Edwin Paul Hubble was an American astronomer. He played a crucial role in establishing the fields of extragalactic astronomy and observational cosmology. He observed and mapped the red-shifting effect of light and postulated that the universe was expanding; 1889-1953 AD

Huggins: Sir William Huggins was an English astronomer best known for his pioneering work in astronomical spectroscopy; 1824-1910 AD

Kant: Immanuel Kant was a German philosopher and one of the central Enlightenment thinkers. Kant's comprehensive and systematic works on metaphysics and aesthetics make him one of the most influential figures in modern Western philosophy. He determined that nebulae were really galaxies; 1724-1804 AD

Lemaitre: George Henri Joseph Edmond Lemaitre was a Belgian Catholic priest, mathematician, astronomer, and professor of physics at the Catholic University of Louvain. He theorized that the universe was created from a singularity or primordial atom; 1894-1966 AD

Ptolemy: Claudius Ptolemy was a mathematician, astronomer, and natural philosopher who wrote several scientific treatises, three of which were of importance to later Byzantine, Islamic, and Western European science. He expanded on Eratosthenes' theories; 100 AD

Glossary of Terms

Big Bang theory: Theory that the universe was started by a small singularity that inflated over the next 13.8 billion years to the universe we now see today.

binding energy: The energy that holds a nucleus together, equal to the mass deflect of the nucleus. In physics and chemistry, binding energy is the smallest amount of energy required to remove a particle from a system of particles into individual parts. The atomic binding energy of an atom is the energy required to disassemble an atom into free electrons and a nucleus.

black holes: A region of space having a gravitational field so intense that no matter or radiation can escape.

carbon dating: The determination of the age or date of organic matter from the relative proportions of the carbon isotopes carbon-12 and carbon-14 that it contains. The ratio between them changes as radioactive carbon-14 decays and is not replaced by exchange with the atmosphere.

cosmic web: A network of filaments of dark matter interlaced with galaxies, believed by many astronomers to form the basis of the universe.

cosmology: The science of the origin and development of the universe.

cosmos: The cosmos is the universe. It denotes the all-encompassing existence of everything throughout an infinite dimension of space.

dark energy: A theoretical repulsive force that counteracts gravity and causes the universe to expand at an accelerating rate.

dark matter: In some cosmological theories, a nonluminous material that is postulated to exist in space and could take any of several forms. It's believed to account for approximately eighty-five percent of the matter in the universe.

Darwinism: The theory of the evolution of the species by natural selection advanced by Charles Darwin.

Einstein ring: Created when light from a galaxy or star passes by a massive object en route to the Earth which distorts the visual image of that object to be seen as being ring shaped.

galactic center: Is the rotational center of a galaxy, which is occupied by massive black holes.

gamma rays: Penetrating electromagnetic radiation of a kind arising from the radioactive decay of atomic nuclei.

gravitational wave: An energy-carrying wave propagating through a gravitational field, produced when a massive body is accelerated or otherwise disturbed. Gravitational waves were first observed directly in September 2015.

gravity well: A gravity well is the pull of gravity that a large body in space exerts.

geocentric universe: Any theory of the structure of the solar system in which Earth is assumed to be at the center of it all.

Great Creation: The theory that God caused the miraculous creation of everything within the cosmos.

heliocentric universe: Is the astronomical model in which the earth and planets revolve around the sun at the center of the universe.

infinite horizon: Viewing the infinite cosmos on a star-filled night from horizon to horizon.

light photon: Visible light is electromagnetic radiation within the portion of the electromagnetic spectrum that can be perceived by the human eye. A light photon is synonymous with a particle of light.

kinetic energy: The energy which a body possesses by virtue of being in motion.

mass defect: The mass deflect of a nucleus is the difference between the total mass of a nucleus and the sum of the masses of all its constituent nucleons.

multiverse: An infinite realm of being or potential being of which the universe is regarded as a part or instance. In other words, a cosmos made up of many universes.

nuclei: Term for the individual particles contained within an atomic nucleus.

nucleon: The plural noun for nucleus.

nucleus of an atom: The atomic nucleus is the small, dense region consisting of protons and neutrons at the center of an atom discovered in 1911 by Ernest Rutherford based on the 1909 Geiger Marsden Gold foil experiment.

neutrino: A neutral subatomic particle with a mass close to zero in half-integral spin, rarely reacting with normal matter. Three kinds of neutrinos are known, associated with the electron, muon, and tau particle.

neutron star: A celestial object of very small radius (typically eighteen miles) and very high density, composed predominantly of closely packed neutrons. Neutron stars are thought to form by the gravitational collapse of the remnant of a massive star after a supernova explosion, provided that the star is insufficiently massive to produce a black hole.

photon: A photon is a type of elementary particle within the electromagnetic field. It has an extremely small mass.

polar shift: A phenomenon that involves the reversal of magnetic poles on earth.

progressive creation: A theory that after the instantaneous creation of everything, God continued to progress His creation through miraculous and continuous interventions.

quantum physics: It's the physics that explains how everything works. The best description we have of the nature of the particles that make up matter and the forces with which they interact.

redshift: A key concept for astronomers. The term can be understood literally—the wavelength of the light is stretched, so the light is seen as "shifted" towards the red part of the light spectrum. Something similar happens to sound waves when a source of sound moves relative to an observer. It also denotes light as traveling slower than the speed of light.

scientist: A person who is studying or has expert knowledge of one or more of the natural or physical sciences.

singularity: A point at which a function takes an infinite value, especially in space-time when matter is infinitely dense, as at the center of a black hole.

speed of light: The speed at which light waves propagate through different materials. The value for the speed of light in a vacuum is now defined as exactly 299,792,458 meters/second.

speed of gravity: The speed of gravity is the rate in meters per second at which gravitational fields or effects propagate through space. According to the classical (Newtonian) physics, the speed of gravity is infinite. The current thinking is that the speed of gravity is equal to the speed of light.

string theory: According to space.com, string theory turns the page on the standard description of the universe by replacing all matter and force particles with just one element. Tiny vibrating strings that twist and turn in complicated ways that, from our perspective, look like particles.

subatomic particle: A particle smaller than an atom. They can be composite particles, such as neutrons and protons, or elementary particles, which according to the standard model are not made up of other particles. Particle physics and nuclear physics study these particles and how they interact.

supernova: A star that suddenly increases greatly in brightness because of a catastrophic explosion that ejects most of its mass.

theory of relativity: The theory of relativity usually encompasses two interrelated theories by Albert Einstein—special relativity and general relativity. Special relativity applies to all physical phenomena in the absence of gravity. General relativity explains the law of gravitation and its relationship to other forces of nature. Einstein's Special Theory of Relativity postulated that $E=mc^2$, which became the most famous equation in the world.

universe: All existing and detectable matter and space considered as a whole. The universe is believed to be at least ten billion light years in diameter and contains a vast number of galaxies.

white dwarf star: A white dwarf is what stars like the sun become after they have exhausted their nuclear fuel near the end of its nuclear burning stage. This type of star expels most of its outer material creating a planetary nebula.

Bibliography

"Timeline of Cosmological Theories". Wikipedia. Accessed May 6, 2020. https://en.m.wikipedia.org/wiki/Timeline_of_cosmological_theories

"Timeline of Discovery of Solar System Planets & Their Moons". Wikipedia. Accessed May 7, 2020. https://en.m.wikipedia.org/wiki/Timeline_of_discovery_of_Solar_System_planets_and_their_moons

"The Fabric of the Cosmos: What is Space?". NOVA, Season 38, Episode 16.

"The Physics of the Universe - Where in the Universe is the Earth?". Accessed May 14, 2020. https://www.physicsoftheuniverse.com/where-in-the-universe-is-the-earth.html

"Dark Matter". Wikipedia. Accessed June 12, 2020. https://en.m.wikipedia.org/wiki/Dark_matter

Redd, Nola. "What is Dark Energy?". May 1, 2013. https://www.space.com/20929-dark-energy.html

"Philosophical Razor". Wikipedia. Accessed May 10, 2020. https://www.google.com/search?client=firefox-b-1-d&q=Wikipedia+%E2%80%93+Philosophical+Razor

"For the Love of Physics – Nuclear Binding Energy". December 21, 2018. https://www.youtube.com/watch?v=BYRz_9wvJzA

Carter, James. "The Living Universe - A New Theory for the Creation of Matter in the Universe". Accessed April 20, 2021. https://living-universe.com/questions-and-answers/how-einstein-was-wrong/

"Gravitational Wave". Wikipedia. Accessed March 2, 2021. https://en.m.wikipedia.org/wiki/Gravitational_wave#History

Siegel, Ethan. "Ask Ethan: Why Do Gravitational Waves Travel Exactly At The Speed Of Light?". Accessed March 2, 2021. https://www.forbes.com/sites/startswithabang/2019/07/06/ask-ethan-why-do-gravitational-waves-travel-exactly-at-the-speed-of-light/?sh=3fedbbab32dc

Siegel, Ethan. "Ask Ethan: Why Did Light Arrive 1.7 Seconds After Gravitational Waves In The Neutron Star Merger?". Accessed March 2, 2021. https://www.forbes.com/sites/startswithabang/2017/10/28/ask-ethan-why-did-light-arrive-1-7-seconds-after-gravitational-waves-in-the-neutron-star-merger/?sh=9ab5c4775d46

Amos, Jonathan. "Gravitational waves: Third detection of deep space warping". BBC Science Correspondent. June 1, 2017. https://www.bbc.com/news/science-environment-40120680

"Georges Lemaitre: Father of the Big Bang". American Museum of Natural History (AMNH). Accessed June 11, 2020. https://www.amnh.org/learn-teach/curriculum-collections/cosmic-horizons-book/georges-lemaitre-big-bang

Bibliography

"The Millennium Simulation Project MPA". Max-Planck-Institute. Accessed February 2, 2021. https://wwwmpa.mpa-garching.mpg.de/galform/virgo/millennium/

Boylan-Kolchin, Michael; Springel, Volker; White, Simon D. M.; Jenkins, Adrian; Lemson, Gerard. "Millennium Simulation-II". Accessed February 2, 2021. https://wwwmpa.mpa-garching.mpg.de/galform/millennium-II/

"History of the Big Bang Theory". Wikipedia. Accessed June 11, 2020. https://en.wikipedia.org/wiki/History_of_the_Big_Bang_theory

"NASA – What Is the Big Bang? NASA Science for Kids". NASA Space Place. Accessed June 11, 2020. https://spaceplace.nasa.gov/big-bang/en/

"Dark Matter". Science Clarified. Accessed June 12, 2020. http://www.scienceclarified.com/Co-Di/Dark-Matter.html

Siegel, Ethan. "If The Universe Is 13.8 Billion Years Old, How Can We See 46 Billion Light Years Away?". May 2, 2018 https://medium.com/starts-with-a-bang/if-the-universe-is-13-8-billion-years-old-how-can-we-see-46-billion-light-years-away-db45212a1cd3

About the Author

JOHN DAYTON

I WAS BORN in 1952 and raised in Southern California. My parents challenged us to seek our full potential, or if you will, reach for the stars. My mother was a wonderful encourager who strengthened my will to do my best. My father was an engineer by profession and a technical manager of hydroelectric power generating facilities in Southern California. He was well respected by his peers for his ability to sort through difficult technical problems. He attained this ability with only a high school education.

I attended local schools where I tested high in spatial perception. I later graduated from California Polytechnic University, San Luis Obispo in 1975 as a Mechanical Engineer with a Nuclear Physics and Engineering concentration.

I worked thirty-six years in the electrical utility industry, most of which was related to engineering and understanding complex technical equipment within the Power Generating Department. I was assigned as a lead engineer at several power plants and became familiar with all types of electrical generation. During this time, I learned how to look at complex problem solving from the perspective of first gathering factual information about the problem, discarding inaccurate information, challenging assumptions, reviewing all of the pertinent facts, and determining the root cause of various problems to derive at workable solutions. So you can see in my case, the apple didn't fall too far from the tree. I've attempted to use these same problem-solving techniques to debate and distill the truth about various subjects in my life.

During my working career, I married Christine, the love of my life, and raised two lovely daughters. But always in the background of my thoughts on many a long commute home from work, I pondered the genesis of the cosmos. As a Christian, I searched for the nexus between science and creation, hoping to provide an explanation that would unite science and religion.

Now that I have retired, I have put my energy into my passion and in the midst of a pandemic have decided to write about what I have uncovered. I believe that my professional life experience and background have enhanced my ability to analyze this subject. I believe that how we grow and learn to question and solve complex ideas is essential to helping us unlock the truth behind solving complex problems and can even help us sort out the origins of the universe.